T0141126

Earth, Ice, Bone, Blood

Earth, Ice, Bone, Blood

Permafrost and Extinction in the Russian Arctic

CHARLOTTE WRIGLEY

University of Minnesota Press
Minneapolis
London

The University of Minnesota Press gratefully acknowledges the generous assistance provided for the publication of this book by the Hamilton P. Traub University Press Fund.

Portions of chapters 2 and 4 are adapted from "Ice and Ivory: The Cryopolitics of Mammoth De-extinction," *Journal of Political Ecology* 28, no. 1 (2021): 782–803. https://doi.org/10.2458/JPE.3030.

Illustrations, unless otherwise specified, were photographed by the author.

Copyright 2023 by the Regents of the University of Minnesota

All rights reserved. No part of this publication may be reproduced, stored in a retrieval system, or transmitted, in any form or by any means, electronic, mechanical, photocopying, recording, or otherwise, without the prior written permission of the publisher.

Published by the University of Minnesota Press
111 Third Avenue South, Suite 290
Minneapolis, MN 55401-2520
http://www.upress.umn.edu

ISBN 978-1-5179-1181-2 (hc)
ISBN 978-1-5179-1182-9 (pb)

A Cataloging-in-Publication record for this book is available from the Library of Congress.

Printed in the United States of America on acid-free paper

The University of Minnesota is an equal-opportunity educator and employer.

30 29 28 27 26 25 24 23 10 9 8 7 6 5 4 3 2 1

Contents

Preface

On the four-hour plane journey from Yakutsk to the Arctic port of Chersky, the landscape changes visibly. The June weather in Yakutsk, the capital of the Russian region of the Sakha Republic, is hot and dusty. Peering through the dirty windows of the Soviet-era Antonov An-24 turboprop plane, I start to notice the ground turn white as we head northward. Then come the mountains—vast stretches of earth crinkled like brain tissue, like balls of paper. As we get closer to the Arctic, the mountains make way for tundra dotted with hundreds of oxbow lakes—something that would have made my high school geography teacher very happy. The plane almost skims the frozen surface of the Kolyma River as it makes its descent toward Chersky's airport. Upon landing, the Russian border guards check the passengers' permits, and we are allowed to disembark. I can feel the "Northness." Up here, the air is fresh and bitingly cold, the sun pale and milky against a hazy sky. There is silence over the Kolyma, punctured only by the scream of a kittiwake and the chattering of families reunited on the runway. The ground is hard, but otherwise there is little to indicate that beneath my feet lies frozen permafrost several hundred meters thick.

A few days later, I accompany the science station's resident botanist, Sergei Davidov, to tend to his "Arctic garden." It's early June and there's nothing there—just frost and frozen soil—but he tells me the names of all the flowers that will push through in the coming weeks. He tells me the climate is changing. As a keen birder, he has noticed a shift in the migration patterns of birds; the pin-tailed snipe now makes its nest on the tundra, when it never used to before. I can scarcely believe that such a cold, gray place will burst into life, but it does: just over a week later the hill is a riot of colors, purple, yellow, blue, and pink, as stubby little flowers have taken root quickly in the thawing permafrost topsoil. Ice from the Arctic Ocean drifts in thick slabs down the Kolyma, carrying with it the trunks of giant pine trees and flooding the lower parts of the town. When I walk into the tundra, the ground is marshy and my boots get saturated with mud and icy water. It never gets dark, and the mosquitoes feast on my blood.

When a large group of American permafrost scientists arrive in Chersky, I go with them to their field sites around the science station. There are a number of controlled-burn sites in the larch forests near Chersky, and the group is here to check on them and see what has colonized the patch of soil since the previous season. It is almost oppressively hot at this point of the Arctic midsummer. The leader of the group, Heather Alexander, cuts into the earth in a neat square shape to produce a permafrost "brownie" that clearly shows the layers of moss, organic matter, and mineral soil that form the permafrost active layer (the part that thaws in the summer). The permafrost in the forest is a little more robust than that on the tundra because of the larch roots that stabilize the ground, but there are patches of waterlogged mire here and there. There's a video (unfortunately, impossible to reproduce in these pages) that shows the team members bouncing on the permafrost as if on a trampoline made of jelly. Throughout my time in Chersky, I witnessed the permafrost earth changing rapidly: freeze to thaw, solid to liquid, continuous to discontinuous.

I was the first scholar from the social sciences and humanities ever to visit the North-East Science Station and the Pleistocene Park in Chersky. I mention this not to blow my own trumpet but rather to highlight both the rarity and the importance of doing critical Arctic fieldwork when so much of what is reported on this region in the global popular media tends toward hyperbole and reinforces stereotypes. Access and ability to conduct this fieldwork is a thorny topic, underscored by inequity in funding opportunities and geopolitical boundary-making practices, much of which privileges Western institutions. The vast majority of visitors to the Pleistocene Park are Western scientists with large grants who are able to afford the $250 a day it costs to stay there and journalists who visit for a few days at a time to write or film short pieces that perpetuate the myth that the Arctic is empty and wild. While my own visit of a month was curtailed by limited funding and access (the station operates only during the summer), becoming immersed in both the practices of permafrost science and the daily rhythms of permafrost living was integral to my understanding of how permafrost as an object is constructed by the different agencies and motivations of those who engage with it. How permafrost is presented is often shallow, barely scratching the surface, treating permafrost as a repository for more interesting things. Digging deeper necessitates standing in the soil, noticing the smaller but no less important ways permafrost affects and is affected by the humans and nonhumans that live with it.

Yakutsk is the largest city in the world built on permafrost, but it is not the only one. The Arctic is a social space, and one that resists the colonial narrative of empty, barren wastelands ripe for exploration and "discovery." Sakha, despite having a population roughly the size of Belgium's spread across a region the size of India, is dotted throughout with towns and villages; the traces of living on and with permafrost are everywhere across the Russian Arctic, but it is important to recognize also how the permafrost *produces* the Sakhan landscape. Yakutsk, once a place of pioneering permafrost science and impressive permafrost infrastructure, now seems skewed. Some buildings sport cracks; some even collapse. The roads are no longer straight, nor are the pipelines, streetlights, and road signs. I wanted to understand what it is like to live in a permafrost city, if only briefly. The Melnikov Permafrost Institute invited me for a stint as a visiting scholar, so I spent the winter of 2018 living and working in Yakutsk, watching the first snows fall in October and the river gradually freeze over into an ice road. Climate change seems a long way off in temperatures that can plummet to minus 50 degrees Celsius, but it makes itself known in the encroaching decay of the city and the lack of funding available to the government of Sakha to make repairs. My adopted colleagues at the MPI have had their research similarly scuppered by poor funding from the Russian Academy of Sciences. I spent my days in their company, learning about their work and their frustrations, becoming aware of the disconnect between the institute's Soviet history and its current state. Their science is different in type from the rather more ad hoc and responsive work done at the science station in Chersky, and it also reveals the discrepancy in funding between Russian permafrost scientists and Western ones. Addressing both issues through an interrogation of how science produces truths and norms is key to understanding the trajectory of permafrost science and the varying responses to climate change–induced permafrost thaw.

I came to Sakha because I wanted to understand permafrost beyond the science, from a humanities perspective. A refrain that I repeat frequently is that permafrost is anything *but* permanent, yet it has long been categorized that way, both linguistically and historically, largely through Soviet science.[1] Just as I watched permafrost change across the summer in Chersky, and across the winter in Yakutsk, its propensity to be slippery, ungraspable, and surprising is the bedrock of this book. Immersion in the permafrost landscape can capture only a snapshot of permafrost life at that current moment. The definition of "living," or indeed "dwelling," is a sticky one, particularly when it comes to conducting research in an unfamiliar place; Tim Ingold describes dwelling as "the immersion of the organism-person in an environment or lifeworld as an inescapable condition of existence."[2] To claim immersion into a landscape or lifeworld as alien to me as Arctic Siberia would not be correct, but an immersive approach to fieldwork means being attuned to the resonances of discomfort and unfamiliarity one might experience as a researcher living, however briefly, in a difficult environment.

It also would not be correct to call what I did ethnography. There is no "ethno" orientation when one is considering an inhuman subjectivity. How, also, to gather data and textually represent an inhuman substance with which one becomes intimately involved, yet without necessarily being aware of it? Elizabeth Povinelli's work on this precise conundrum rejects Clifford Geertz's now-famous interpretation of ethnography as "thick description," which asserts that the objective of the ethnographer is to extract meaning from the culture being studied.[3] Povinelli, in thinking through her work with Australian Aboriginal communities, prefers to situate herself in "thin description." By tracking a sore on her own body and drawing attention to the ways in which sores are materially and socially produced through a colonial regime, she refuses the presumption that an objective truth might emerge from the researcher. She states: "I am especially interested in an aspect of social life that I am calling carnality: the socially built space between flesh and environment."[4] This materiality, both bodily and grounded, produces a dynamism that shapes social life as much as cultural representation and allows for the unpredictable agencies of the inhuman to reveal themselves. I track my own body as it moves through the permafrost environment.

In his book *Stone: An Ecology of the Inhuman,* Jeffrey Jerome Cohen claims: "The stories we know of stone will always be human stories, even if the cosmos they convey makes a problem of that category rather than celebrates some specious natural dominion."[5] This is the fundamental problem of writing about an inhuman substance such as permafrost, but also an opportunity to practice a different sort of storytelling that is responsive and generative. Immersive fieldwork that takes matter as its primary interrogation should be attuned to, and represent, the hybridity of its methods, which can never presume to know the material substance, but rather should seek to reveal inhuman subjectivity and its relations with other actors—both human and nonhuman. This is a difficult task, particularly when one is attempting to write about grander scales than that of human social worlds or to speak to inscrutable and shifting material processes. My methods included, of course, many interviews and conversations with human actors who engage with permafrost, whose stories I relate here, but I also tell the stories of nonhumans (the reindeer who make their home on tundra, the animals brought to the Pleistocene Park in trucks and boats) and recount the dynamic materiality of permafrost at a time of climatic upheaval in the Russian Arctic.

Throughout this book I emphasize the material process of discontinuity—that of permafrost thaw and its reverberations. I conducted this research with a commitment to openness, to being responsive to the changes in the permafrost landscape and how these affect the humans and nonhumans who live there, but I brought with me my own particular background and training. I am a human geographer who situates herself within the environmental humanities or, even more specifically, the Arctic humanities.[6] In this book, I aim to attend to both the spatialities and the materialities encompassed by the Arctic and its permafrost landscape; this is an interdisciplinary text, but it is still bound by academic norms. Practicing discontinuity demands that I recognize my own limitations as both a researcher and a storyteller, and acknowledge that I can construct only part of the picture. As I consider how to tell stories of the permafrost that do not lose sight of its creativity and surprise, I remind myself of the discontinuity I inhabit by entering and leaving the field site, of the stories I pick up along the way but also the stories that I leave behind.

The stories I tell throughout these pages emphasize the specificity of permafrost thaw to life in Sakha, but they are also saturated with wider implications for the planet. In this book, I gather together instances of permafrost thaw and its knock-on effects as a way to reveal the heterogeneity of permafrost as a substance but also as a meditation on the meaning of loss and decay in a region that has evolved with the rhythms of permafrost for centuries. Now those rhythms are out of sync, the gradual buildup of knowledge and understanding of *how* to live with permafrost becoming redundant in the face of anthropogenic climate change and Arctic heating. How to conceptualize this? At a time of massive environmental upheaval and destruction, how to understand just what is being lost and who exactly is losing it? One word kept coming up again and again: extinction. Through my fieldwork at the Pleistocene Park and in Yakutsk, it was impossible to ignore the undercurrent of restoration, revival, and resurrection that a thawing permafrost generated, whether that be overtly in the case of mammoth de-extinction or more abstractly through the restoration of the Pleistocene ecosystem in Northern Sakha. The specter of extinction is everywhere, and as the planet enters its sixth great extinction period (the first one anthropogenically caused), it becomes necessary to interrogate the normative definition of extinction as pertaining merely to species. Opening up the definition of extinction to encompass ways of life, knowledges, cultures, inhuman substances (ice, permafrost) and processes, histories, and temporalities reveals a less tangible but no less devastating sense of loss.

I wrote this book to demonstrate how permafrost can act as a conduit to understanding extinction in the Anthropocene while staying rooted in place—simultaneously a vast region of global significance and a microscopic crystalline substance. Permafrost also embodies the meaning of coldness in a world heating rapidly, inextricably linked to life and survival as the Arctic melts. Who gets to remain cold (and thus survive) is a question of power and politics, not merely spatial arrangements across the planet. The climate crisis cannot be addressed without acknowledgment of the destructive environmental practices that have created it in the first place, and that acknowledgment in turn requires recognition that such practices are carried out by only a small percentage of humans, along with a capitalist system built on resource domination. What can be seen happening all across the global North is an obfuscation of the perpetrators of ecological violence—a greenwashing tactic purporting that there is no need for them to change their destructive behavior because technological ingenuity will fix everything and mastery over planetary processes is the goal of living in the Anthropocene. This book challenges that idea, reorienting the normative definition of extinction through the discontinuity of permafrost and how it refuses the various attempts to bring it into the realm of the powerful.

Introduction

Permafrost Life

What is permafrost? This seemingly innocuous question has a surprisingly complex and enigmatic series of answers. Prying the word open, it appears to be a portmanteau of *permanent* and *frost,* but it is not that simple. Frost can take many forms, most usually as an accumulation of ice on a surface, but it is decidedly not permafrost. Indeed, in the mid-twentieth century the name itself was the subject of intense debate between the United States and the Soviet Union, and even between factions within the Soviet Union. The so-called father of permafrost, Mikhail Sumgin, emerged from this fight victorious, while his rival, Sergey Parkhomenko, faded into obscurity—the prize of posterity meaning that Soviet science subsequently defined permafrost as a physical object that is quantifiable on a map, in contrast to the more abstract "processual" definition favored by Parkhomenko.[1] When it came to translating the Russian term, *vechnaya merzlota,* the English language was unable to do the word *vechnaya* justice, and scientists settled on *permanent,* as opposed to a word that might convey the rather more vague concept of continuity over time that the Russian word captures.[2] Although debates continued (and still occur) regarding permafrost's status as either a physical "thing" or a more dynamic process defined by temperature, the term *permafrost* stuck, and in its simplest form refers to permanently frozen ground.[3] It is this definition I work with, and subsequently problematize, in this book. The main problem, of course, is that permafrost is anything but permanent.

Permafrost is the iceberg's less flashy, less noticed cousin. The iceberg is gleaming and pristine, home to the majesty of polar bears and balancing serenely on frosty-blue water. It is the image that encapsulates global warming, the ice disintegrating visibly into the ocean and melting under an unforgiving sun. Permafrost, by contrast, is dull and marshy. It is largely hidden under snow for much of the year, emerging in the brief summer months infested with mosquitoes, fetid and humped. Environmental science defines permafrost as subsurface cryotic material—usually soil—that maintains a constant temperature of zero degrees or below for two years or more, with a surface "active layer" that thaws and freezes seasonally.[4] Geographically, it covers a quarter of the Northern Hemisphere's land surface, and, by trapping the greenhouse gases carbon dioxide and methane, it sequesters more than double the amount of carbon currently found in the atmosphere.[5] Scientists who have their eyes trained on this broad latitude of land point to worrying trends and unpredictable events, such as the opening up of huge sinkholes—termed *thermokarst megaslumps*—which occur when accelerated thawing exposes large surface areas and the ground collapses in on itself. Methane gas bubbles explode, forming craters that dot the landscape. Drunken forests are caused by uneven thaw; house foundations sink into the mud while walls crack; roads buckle and power lines break; coastlines slump into the sea; carbon-rich yedoma begins to disintegrate.[6] In its 2019 report on the cryosphere, the Intergovernmental Panel on Climate Change (IPCC) predicted with a high degree of confidence that widespread disappearance of permafrost will occur with climate change, and this will produce major knock-on effects; there is a growing fear that the greenhouse gases escaping in small quantities now will create a feedback effect of faster and more widespread warming, triggering the release of even more methane and carbon.[7] The permafrost landscape is visibly changing, and the Arctic, warming at three times the rate of the rest of the planet, makes the global news with increasing regularity.[8] Wildfires, record temperatures, an increasingly ice-free Arctic ocean—barely a day goes by without some form of worrying occurrence emanating from the northernmost parts of the world. Historically so often forgotten or misunderstood, the Arctic has now become a climate barometer for the rest of the planet. With 65 percent of Russia's landmass comprising permafrost, and one-fifth of it above the Arctic Circle, the history of Russia and the Soviet Union's engagement with Arctic landscapes is unique, a history that curates the particular permafrost milieu of Russia and reverberates into its fraught present and future.

The Russian Arctic and Remoteness

In her beautiful ode to remoteness "The Lena Is Worthy of Baikal," Ksenia Tatarchenko reveals the intimate connections and attachments to place in what has often been conceived of as a "nonplace." Rather than painting the Arctic as a harsh and unforgiving landscape, devoid of features or people, Tatarchenko presents a careful portrait of the Russian (or Soviet, in this case) North that points to a more nuanced history, dismantling nature/culture binaries and interrogating the effects of a Soviet politics on the Arctic today. The trope of the (white, male) Arctic explorer is subverted neatly by the 1973 Soviet film adaptation of permafrost scientist Vladimir Obruchev's 1924 sci-fi novel *Sannikov Land,* in which a group of sailors hunt for a phantom Arctic island. Their eventual success brings destruction not only to themselves but also to the Indigenous population, and the film closes with their echoing refrain of "Man, why are you marching on the Earth?" Tatarchenko draws parallels between the film's depiction of a disintegrating Arctic and the current climatic disruption as the catalyst for novel engagements and connections with changing (melting, thawing) Arctic landscapes, likening the groups of Arctic researchers drawn to these remote locations to "tribes," their practices of knowledge production "differentiated by experience and authority and where acceptance is a matter of rituals." Permafrost's difficult geography curates the coalescence of locals and researchers, of history and present, of isolation and global connection, and while a map might trace a rich vein of Arctic settlements across Russia, there is no escaping the material reality for these towns, of being remote in a warming world—they are, as Tatarchenko states, "a monument to the violence of abandonment."[9]

This has not always been the case. The abandonment Tatarchenko identifies is a particularly devastating compound of political upheaval, decimated state provision, and the painful bite of unfettered capitalism that flooded Russia in the decades after perestroika.

During this period in the Soviet Union, thousands of Russians were lured to the Arctic from densely populated areas in the country's western regions with the promise of well-paying jobs and housing as part of a state plan to "master the North." It made sense: one-fifth of Russia's huge landmass is north of the Arctic Circle, and 65 percent of it is permafrost in varying forms. As towns and cities grew, the knowledge of how to build on top of permafrost grew with them.[10] Obruchev himself was, alongside Mikhail Sumgin, integral to the nascent discipline of permafrost science and engineering. Despite the difficult climate and remoteness of Arctic life, the population grew rapidly as a result of the incentives offered by the state. The utopia promised, however, corresponded only to a particular Soviet fantasy. Arctic Indigenous peoples, particularly nomadic reindeer herders, were forcibly assimilated into a homogeneous "Soviet culture," separated from their belief systems and ways of life. Indigenous languages and traditional knowledges were decimated, and quotas were placed on reindeer farming, which now took place in specially built collective settlements. The production of this Soviet culture was intimately entangled with fantasies around Siberia and the Russian Arctic as a hostile space to be conquered—the heroic and "civilizing" pursuit of communism through technoscientific progress put into practice.[11]

Then perestroika happened, the Soviet project collapsed, and Russia was plunged into economic turmoil; while mafia groups proliferated and new oligarchs thrived, the people who made their homes in the Arctic were largely left to fend for themselves as centralized state provision shrank and Moscow seemed to retreat farther and farther away. Since the early 1990s, the rate of out-migration from the Russian Arctic has reached almost 75 percent in places, while jobs are scarce and the cost of living is eye-wateringly high.[12] Meanwhile, Russia has ramped up its economic interests in the Arctic in recent years, and its current Arctic strategy involves a series of ambitious—and destructive—developments for the region that seem to reveal a renewed interest in mastering the North, but without much of the utopic incentive that accompanied the Soviet plan. These developments include mines that bank on the increased thawing of permafrost to "unlock" the resources within, a bloated extractive industry focusing on oil and liquefied natural gas production, and an Arctic Ocean shipping route between Europe and Asia that is 40 percent shorter than the current shortest route, via the Suez Canal.[13] The seventeen-page document that lays out Russia's Arctic strategy makes little mention of the negative effects climate change will have on these proposals; instead, it plays up the economic benefits of a warming Arctic and melting sea ice while conveniently ignoring the crumbling infrastructure and greater likelihood of floods and fires that will accompany an increasingly unmanageable climate.[14] This is Russia in the Anthropocene.

The permafrost underscores all these disparate configurations of life in the Russian Arctic and is absolutely inextricable from them: as a foundation for a utopian building project or the shallow grave of a totalitarian regime; as a fantasy for colonial explorer-scientists or a home world for Indigenous Siberians; as a prospective cache of valuable resources or a repository of escaping carbon. As permafrost thaws, the different relations produced in these spaces of absence and retreat take on greater significance as nodes of survival on an increasingly unstable landscape. It is this instability, much of it a result of anthropogenic climate change, that has led some geologists and other scholars to assert that we are now living in the Anthropocene—a new geological epoch that has supplanted the relative stability of the Holocene and that designates geological and ecological dominance to humans as a species.[15] The realization that permafrost's planetary cooling effect is, like the Arctic in general, integral to human life is indicative not only of the responsibility bestowed by the Anthropocene moniker but also of an existential dread that strikes at the heart of an apocalyptic narrative that has been growing in Western cultural media and news reporting. We are officially living in the time of the sixth great extinction—a period during which the background rate of species extinction is much higher than normal—and the first of Earth's extinction events to be anthropogenically caused.[16] The planet is losing creatures and other forms of life at a headlong rate. Meanwhile, the thought of runaway climate change, of "tipping points" found within permafrost and Arctic sea ice, forces a level of introspection that ends in an unanswerable question: Are we next?

Extinction is not a new idea, yet as the Anthropocene lays bare both the fragility of life on the planet and the potential power humans might wield over it, it becomes necessary to demand new ways of looking at extinction, interdependence, and responsibility. This requires stepping outside of the notion that "we," as humans, are somehow separate from the rest of nature; it requires thinking of species beyond our own, and beyond species categorization in general. Can tundra go extinct? Can the Arctic die? And is it possible, indeed, to conserve or resurrect permafrost? Over the course of this book, I will revisit this question again and again, unraveling the knots of life and nonlife found atop (and within) a permafrost landscape that is shifting, retreating, disappearing—but also one that is subject to various revitalizing attempts to slow and turn back time, to curate a safe space for human (and nonhuman) survival. Permafrost frequently confounds these attempts by being uncontrollable and unpredictable; it is within this sense of refusal that the paradox of the Anthropocene lies—we are now the dominant planetary species, but that existence is highly dependent on the Earth *and* the earth beneath our feet.

What binds these anxieties of survival is coldness. The knowledge that the planet will very likely continue heating to a state of uninhabitability has not seemed to inspire any meaningful change in behavior from the world's biggest polluters and destroyers, but there has at least begun to be an understanding that life and temperature are inextricably linked. Thinking about coldness and life through different modes of power corresponds to what some scholars are calling "cryopolitics"—a riff on Foucault's theory of biopolitics that aims to draw attention to the novel ways freezing and thawing reconceptualize life through preservation, extension, and storage. Foucault's definition of biopolitics as "make live and let die" is cryopolitically altered to produce "a zone of existence where beings are made to live and are *not allowed to die*."[17] This seemingly minor tweak has major implications for the control of, and over, life; freezing implies a temporal stasis of both life and death, as well as the potential to suspend (and, indeed, reverse) extinction. Decisions about which lives get to experience, to echo the Indigenous activist Sheila Watt-Cloutier's phrase, "the right to be cold" are buttressed by the caveats of power, money, race, and gender as entangled categories of climatic and geographic privilege.[18] The people (and animals) who live in the Arctic are affected differently by coldness and the climate crisis than those who live elsewhere, and the stories I tell throughout this book reflect the specificity of permafrost life as inextricable from Russia's history and approach to politics, encapsulated particularly within one permafrost thaw mitigation scheme being played out in an isolated area of the Russian Arctic.

A Mammoth Solution

In a remote corner of the Sakha Republic, in northeastern Russia, a unique project has been under way since the 1980s. This is the Pleistocene Park, where the father and son owners are attempting to address the potential catastrophic outcomes of permafrost thaw through ecosystem rewilding. Sergey Zimov, a geophysicist, set up the North-East Science Station (NESS) in 1977 in the Arctic town of Chersky. From the station, he monitored shifts in the carbon cycle, methane fluxes, and paleoclimatic trends, and he began to notice a link between permafrost soils and carbon sequestration—something that had rarely been addressed in permafrost science before. To prove his hypothesis, he conducted an experiment in which he cordoned off an area of nearby tundra to see what would happen to the permafrost if he began to restore the ecosystem that had existed there thousands of years ago. Populated by megafauna such as mammoths and woolly rhinoceroses, alongside a vast array of insects and other creatures, this ecosystem looked very different from the tundra of today. Where now there are stunted larch trees, shrubs, and very low biodiversity, the Pleistocene landscape was a grassy steppe that supported and was supported by many forms of life—or, as Sergey puts it, a vast "Arctic savanna."[19] It is this landscape that Sergey—now with his son Nikita on board—is attempting to re-create, to restore the benefits of such an ecosystem for the permafrost: in the Pleistocene, gigantic beasts trampled and compacted the permafrost, keeping it frozen, and the light-colored grassland stimulated by animal grazing created an albedo effect that reflected much of the sun's hot glare back into the atmosphere. Currently occupying a territory of two thousand hectares surrounded by a fence, the Pleistocene Park has slowly expanded its collection of animals and is now home to a variety of large herbivores: musk oxen, wild horses, reindeer, bison, goats, camels, sheep, Kalmykian cows, yaks, and moose.[20] These animals are tasked with a weighty labor: to stop the permafrost from thawing and the carbon escaping. Eventually, the Zimovs envisage a network of similar parks across the entire Arctic permafrost landscape.

Despite the somewhat fantastical nature of their project, the Zimovs are pragmatic scientists who have spent many years living on the tundra (although they now spend the winters in more hospitable climates). Their science is underpinned by the daily rhythms of permafrost life—the vagaries of maintaining equipment and keeping their animals alive. Receiving no government support and considered too maverick by the Russian Academy of Sciences to qualify for funding, they largely pay for the park by charging fees to host international permafrost researchers every summer at NESS, which is housed in a repurposed Soviet television station. Hundreds of miles away in Yakutsk, the capital of Sakha, the Melnikov Permafrost Institute sits as a fading pioneer of Soviet permafrost science, while the city itself—the largest built on top of permafrost—is an incredible feat of engineering. Today, the MPI suffers from a lack of funding, while Yakutsk's infrastructure buckles and warps as the permafrost thaws. Beyond the boundaries of the city are Sakhan villages where the residents breed cattle and horses, and in areas even farther out live Indigenous reindeer herders, some of whom have managed to resist Soviet collectivization and remain nomadic, and some of whom live in settlements that dot the Kolyma basin. These groups know their *alaas* feeding grounds are being destroyed by thermokarst; they know that unseasonable warmth brings rain instead of snow, which freezes quickly into ice, so the reindeer cannot feed on the foliage underneath.[21] And their reindeer are being infected with decades-old strains of anthrax, which escape from the mushy ground during particularly warm summers.[22] The Pleistocene Park fits, albeit uneasily, alongside these other ways of knowing and living with permafrost.

The Anthropocene is played out in a variety of ways in this small, isolated corner of the globe. It is Sergey's belief that humans wiped out most of the Pleistocene megafauna through hunting overkill and created the barren tundra that exists today—a controversial hypothesis disputed by many who argue that climate fluctuations were to blame.[23] The idea of "resurrecting" the mammoth's ecosystem came from the discovery of thousands upon thousands of Pleistocene animal bones—mammoth, woolly rhino, bison, horse, deer—in a permafrost bank on the Kolyma called Duvanny Yar. In contrast to the empty tundra that currently exists, where the most abundant creature is the mosquito, these bones suggested a teeming ecosystem that, whether you believe the cause was Pleistocene overkill or climate, has now disappeared. These bones—and in some cases whole bodies—have been preserved by permafrost for centuries: a mausoleum of multiple extinctions, some planetary, some local, some recorded in history books, some lost to history long ago.

The Zimovs' attempt to reverse this process is part of a wider series of rewilding experiments across the planet that follow a holistic conservation process: instead of focusing on saving individual creatures from extinction, these projects encourage entire ecosystems to flourish through the reintroduction of keystone species and an emphasis on natural processes with little human intervention. The animals at the Pleistocene Park are confined to the fenced-in area and occasionally monitored by park rangers, but aside from that they roam as they please. Plucked from Mongolia, China, other parts of Russia, and even Denmark, they have crossed thousands of miles in creaking trucks, traversing ice roads and fording rivers. They have been taken by boat from Wrangel Island in the Arctic Ocean, one of the very last strongholds of the mammoth before its extinction. There was even a scheme, before it collapsed due to lack of a pilot, to charter a plane from Alaska to transport a herd of sedated baby bison to the remote airfield in Chersky. All this to re-create the prehistoric mammoth steppe ecosystem and, perhaps, to save the world.

This is the Pleistocene Park's plan. It is not about protecting animals, or even about revitalizing ecosystems; it is a geoengineering strategy to protect humans from catastrophic permafrost thaw and the devastating climate ramifications this may have.[24] The doomsday narrative found within the park's ethos and approach is very much one of potential human extinction, yet it is inextricably entangled with other forms of extinction, be they past, future, or ongoing. Nowhere is this more apparent than in the extinction of the woolly mammoth, the corpses of which are exposed on the tundra in ever-increasing numbers as permafrost retreats. Bringing this extinct creature back to life has become the ultimate goal for the controversial new science of de-extinction, which purports to be able to accomplish that goal through gene technology—either through cloning or hybridization. While *Jurassic Park*–style dinosaurs are beyond the capacity of de-extinction science (the creatures are just *too* dead), mammoths are not; having lain in their permafrost graves for centuries, many mammoth bodies have been frozen in time, almost as if sleeping, with their DNA kept nearly fully intact by the preservational properties of ice. While the Pleistocene Park is not directly involved with mammoth de-extinction (although it is certainly tangentially involved), it stakes its claim to any successful resurrections by way of restoring the mammoth's former habitat, and through its name's tongue-in-cheek reference to pop culture's most notorious de-extinction project.

The woolly mammoth plays a role in the mythologies of some of the Indigenous groups who live on the tundra: Sakhans believe the mammoth to be a giant ratlike creature that tunnels underground, creating permafrost features such as polygons, while the Evenki consider the mammoth to be aquatic and believe that the reason no mammoth has ever been found alive is that any exposure aboveground will kill them.[25] That so many mammoths are now found at the surface, carcasses thawing and rotting, can only be a sign that something is up down below. Western science places the mammoth's extinction at around ten thousand years ago, although small populations survived on St. Paul Island near Alaska until 3750 B.C. and on Wrangel Island until 1650 B.C. Many of these mammoths would have fallen into swamps or mudbanks, which then froze over, preventing decomposition from occurring; some scientists have estimated that up to ten million mammoths may have been preserved in permafrost this way.[26] Now they are thawing. Savvy Sakhans have noticed the abundance of mammoth bodies with their ivory tusks intact, and a thriving black-market industry in tusk hunting has emerged, with the majority of tusks sold to China, where they become elaborately carved status ornaments. Survival is hard for the people who live on permafrost, from the city to the Arctic tundra, with temperatures plummeting to minus 60 degrees Celsius in the winter and the sun barely rising above the horizon for several months of the year. In an Arctic where jobs are hard to come by and the price of living is astronomical, spending the summer season hunting for mammoth carcasses can mean another year of relatively decent pay for the tusk hunters and their families, but there is no telling how long this bounty will last; no one knows for sure just how many of these gently rotting prehistoric beasts the landscape holds.

The different manifestations of permafrost in the Anthropocene are spread across time, space, and scale. They take the form of scientific knowledge, of hypothesis, of apocalypse, of story and myth, of legal representation, as preserver of history and destroyer of futures, rooted in place while also stretching across the entire Northern Hemisphere, as a delineation on a map or a reading on an instrument, the feeding ground of a Kalmykian cow and the resting ground of a long-dead mammoth god. Permafrost is a Sakhan freezer for milk and meat, a storage facility for permafrost cores, a repository of ancient viruses, and a catacomb of Pleistocene bones; it is also a huge carbon sink and a microscopic grain of earth. Permafrost produces, alters, and destroys different configurations of life and death, renegotiates the boundaries of survival, and redefines what extinction means. Attempting to separate all these things will not work, nor will trying to gather them all together. They are at once one thing and another, separated but inseparable. They are, in a word, discontinuous.

Permafrost and Extinction

The scientific definition of permafrost categorizes its geographical spread as either continuous or discontinuous, with continuous deep permafrost being found in the highest latitudes and discontinuous permafrost found farther south.[27] Some areas of continuous permafrost are centuries old, formed during the glacial period predating the Holocene epoch and surviving because of the reliably cold winters that marked the Arctic region until recently; discontinuous permafrost is less stable, subject to more seasonal temperature fluctuations.[28] Discontinuity is creeping northward, however; in 2018, Nikita and Sergey conducted initial experiments on the continuous permafrost around the Pleistocene Park and discovered that some of it was failing to freeze again in the winter.[29] The consequences and implications of this finding are enormous, not only for the permafrost landscape and climate but also for all forms of life on the planet. Discontinuity is a material process underscored by thaw and a warming Arctic, but it is also a mode of thought and being concretely rendered by the uneven keel of living—and surviving—in the Anthropocene.

Throughout this book, I explore the idea of discontinuity, proposing it as a reorientation to normative thinking around extinction in the Anthropocene: thinking that defines the "Anthropos" as a homogeneous category, that sanctions the destructive practices of the extraction industry (and its subsequent climatic technofixes), that renders life controllable and manipulable. To think discontinuously requires becoming attuned to the slippages and jagged discomforts of planetary life and death on, and in, the ground, but it also involves imagining what might disrupt—*discontinue*—that which came before. What are the new relations that might form in this unpredictable space? How can the modernist dichotomy of nature versus culture be dismantled? And how does permafrost figure in this, as both a barometer for earthly survival and a subterranean, indifferent landmass? Discontinuous thinking necessitates an interdisciplinary engagement with theory and knowledge, with the understanding that it is always an ongoing process; as the permafrost changes, the relations and knowledges that emerge from these shifts change also. Living and dying in the Anthropocene both informs and is informed by permafrost freeze and thaw to the extent that the process of becoming extinct (or, indeed, de-extincted) is in constant flux.[30] Extinction is no longer just the end of a species, something that is coded into life itself; it is the fundamental paradox of an epoch that designates humans as geological masters on an increasingly restless planet that threatens all earthly survival.[31] My aim in this book is to extend current thinking around extinction by using permafrost and its concomitant processes of *becoming* discontinuous to reimagine modes of life, death, and survival that are not enclosed events, but rather an intersecting web of different temporalities, scales, materialities, and ontologies. By rooting my discussion in the Russian permafrost landscape—specifically, Sakha—I point to the importance of historically contingent and local understandings of permafrost life and demonstrate that homogeneous definitions of life and extinction cannot simply be scaled up or mapped elsewhere.

Extinction as an Idea

Claire Colebrook identifies three different types (or "senses," as she calls them) of extinction in the Anthropocene: the sixth great extinction event, extinction of individual species by humans, and human extinction itself.[32] These categories formalize extinctions as events, either future or ongoing, contextualized by an anthropogenic global warming that produces particular anxieties and fears for planetary survival; the points at which these fears crystallize are what produce the idea of an extinction event. Extinction is, however, much more complicated than that. The event suggests there is an element of quickness to extinction—the day of reckoning, a big bang—but in reality, most extinctions are long, drawn-out processes that might not even register on human quantitative methods such as those used to create so-called red lists.[33] Thom van Dooren conceptualizes extinction as a "slow unraveling," an often imperceptible process of the entangled relations of species life coming apart and fraying at the seams.[34] The ethical implications of this are enormous. To be the cause of that, as humans are in the sixth great extinction, necessitates an introspective engagement with both the responsibility toward multispecies kin in the Anthropocene and the ways in which these relations are coproduced—how they encompass, as Deborah Bird Rose puts it, "multispecies knots of ethical time."[35]

How do these particular modes of Anthropocene extinction square with the Darwinian notion that extinction is coded into species life? As Cary Wolfe puts it: "Extinction both is the most natural thing in the world and, at the same time, is not and never could be natural."[36] Surely the fundamental premise of the Anthropocene, not to mention several decades of careful scholarship in the sciences and humanities, has by now rendered defunct any sort of attachment to the idea of a natural sphere separated from human activity? To add to this, the notion of species categorization itself is subject to scrutiny—species are rendered legible in all sorts of ways that ignore the heterogeneity found within species groups, and similarly that legibility is confounded in all sorts of ways too.[37] How a species might be conceived of in a textbook is different from what it becomes on a red list, not to mention the myriad subversions made by lively individuals that officially belong to that species—hybridizing, transgressing territory, evading population counts. Nevertheless, that death is coded into life is a given; surely there might be some merit in the notion that extinction is an *extension* of the species, given that the planet has already experienced five other mass extinctions? The paradox of the Anthropocene rears its head once more as we grapple with the mantle of responsibility for the sixth, yet at the same time are forced to stare down the barrel of our own, self-administered demise.

Extinction suggests finality; it promises ends and no returns. Thomas Moynihan identifies a historical shift from the eschatological apocalypse of religion to a secular extinction narrative that does away with the notion of redemption or afterlife: extinction in modernity is the end of the line, and that is a terrifying prospect.[38] Yet extinction is not the end of all ends. The hubris of imagining the end of humanity as the end of time itself ignores the reality that there is an afterlife to extinction.[39] What might be found in the absences left by extinction? These are undoubtedly hubristic times; tales of nonhuman extinction tend to focus on the heroic exploits of conservationists or on guilty elegies of loss.[40] Regardless of what form they take, they are stories about *us*.[41] Meanwhile, imaginaries of human extinction—dominated by Western media and Hollywood—usually depict the world being saved in the nick of time, or, if disaster occurs, as a postapocalyptic wasteland inhabited by pockets of humans attempting to kick-start “civilization” again. Stories are important for making sense of extinction, but who is telling the story, and how it is being told, matters hugely. Such stories can make legible how extinction is less a straight trajectory of decline and more a process of unraveling and remaking life as economic and political conditions structure the possibilities for existence.[42] What I wish to encourage with a discontinuous approach is the idea that extinction does not have to be a moralizing tale that situates the human subject as the protagonist, nor does it have to be the end of the story. Discontinuity imagines a way out of the linear narrative of life to death to extinction and nothingness, instead allowing for something more generative to emerge from thinking through extinction differently.

In this book I tell multiple stories of Anthropocene extinction, all of them contextualized by permafrost in some way. If extinction is understood to be a heterogeneous and slow unraveling of time and relations, how can storying make explicit something that is often difficult to know, sense, and describe? How do stories account for both devastating ruptures and the more hidden slippages of life and death? And how does permafrost make itself known *through* these stories as a protagonist, rather than a backdrop? Like Colebrook, I identify different forms of extinction; each of the four chapters of this book addresses one of these forms, although I acknowledge that these are certainly not the only ones. In the first chapter, "Earth," I tackle the underlying fear that is coded into Anthropocene living: that of human extinction, and how a turn toward apocalyptic narratives obscures smaller and ongoing apocalypses that disrupt the idea of a catastrophic future event. In "Ice," I consider more intangible and unrecorded forms of extinction by thinking about permafrost thaw and what this means for different ontological configurations of ice and tundra. In "Bone," I think through species death and how extinctions are rarely linear processes by showcasing the temporal stickiness found in ecosystem rewilding. Finally, in "Blood," I take the idea of de-extinction and consider the implications of mammoth resurrection as a permafrost mitigation strategy designed to save the world—and the humans in it. The chapter titles are meant only to loosely anchor the subject matter and should be taken as starting points from which to unsettle normative categories of extinction, underscored by the thawing of permafrost.

Human Extinction

"What other Earth-born species can think upon its own demise?" asks Thomas Moynihan in his book *X-Risk,* which tracks humankind's preoccupation with its own extinction across history.[43]

Humans have imagined the apocalypse, the Rapture, the end times in myriad ways throughout history; prophecies, cults, and entire religions have been built around the idea that one day we, as a species, will cease to exist. In the Anthropocene, however, the likelihood and severity of our extinction becomes truncated: as (some) humans exert greater power over earthly processes by mining, extracting, and polluting, those processes tend to become less optimal for survival.[44] A growing awareness of the impacts of climate change, as well as the material effects of environmental destruction—hollowed-out toxic mines, the great plastic island in the Southern Ocean, the bleaching of the coral reefs—has resulted in a shift in attitudes toward how humans engage with the natural world. What is known as the modern Western environmental movement emerged in the United States after World War II, contextualized by technoscientific progress spurred on by the Cold War and a growing understanding that human actions can influence ecological processes. The famous "pale blue dot" photograph of planet Earth seen in its entirety from space for the first time produced an understanding of ecology that posited human (and nonhuman) life as global, interconnected, and fragile.[45] Stewart Brand's "whole Earth" doctrine and similar movements curated a utopic narrative of the eco-warrior fighting for a better planet for all, with the implicit understanding that humans are obligated to provide benign stewardship for the natural world (a position that Brand maintains to this day and that drives his de-extinction efforts).[46] But as scientists began to develop a better understanding of anthropogenic climate change and biodiversity loss, the utopia imagined by the original environmentalists shifted to more of a dystopia. With new paleontological revelations that attributed the extinction of the dinosaurs to an asteroid rather than evolution, alongside the political upheavals of the 1970s and 1980s, human existentialism and apocalyptic narratives took on a more nihilistic bent that suggested the fate of human life on Earth was out of our hands.[47] Thus the paradox of the Anthropocene creates the ultimate dilemma: whether to use humanity's status as dominant geological agent to further steward the planet or to step back from causing further harm.

The crux of the Anthropocene is, first and foremost, one of scale.

At first blush it might seem prescient to attribute to all of humanity the power of a major geological force, but as a globally defined epoch, the Anthropocene requires that we consider the *entire* global population (and its ancestry) as both signifier of the sort of geological agency needed to force an epochal shift and receiver of whatever conditions emerge from such agency. The Anthropocene concept necessarily also scales up extinction discourse: the sixth great extinction encompasses the whole planet, and so it stands to reason that humans become subject to the same globalizing processes. The specter of apocalypse emerges from the fear that modernity and its concomitant assumption of human exceptionalism are coming to an end.[48] This is not to say that humans routinely imagine their future extinction, no matter how bad the news gets (and humans, incidentally, fall into the "least concern" category on the International Union for Conservation of Nature's red list), but the idea takes on greater weight in the face of the anthropogenic climate crisis.[49] We are now an epoch-defining species, and that demands a certain introspection into what species-being actually entails. Dipesh Chakrabarty argues that this involves a need to decolonize the term—as empire builder and colonizer, the West has claimed its history as a sort of universal truth, obscuring the histories and voices of the colonized, and has become synonymous with humanity as a whole.[50] But species thinking in a time of capitalist globalization and Western hubris is bound to become mired in universalist tropes, which must be overturned through attention to scale.[51] The perpetrators of Anthropocene dominance and destruction are not the same people who reap the disastrous effects.[52] To speak of extinction—the end of humanity—is a privilege afforded to those for whom such a possibility is a long way off, or even avoidable if one happens to be rich enough.[53] For those people whose lives are already threatened by climate change, extinction might take the form of smaller, but no less devastating, events.

"To periodize the Anthropocene is already to assume a future world in which human presence on Earth has been reduced to a lithic layer," states Richard Grusin.[54] This viewpoint assumes that the Anthropocene will end only when humans as a species do (a viewpoint I am reluctant to share, as it leaves little room for imagining a better world) and places the human extinction event squarely in the future.

Indeed, the purely prescriptive definition of extinction—that every last human on Earth will die—validates this statement. But when we begin to draw on other temporalities and scales, a different picture emerges, one that considers extinction as a heterogeneous patchwork that does not merely coalesce at a single point *in* time but also is arranged discontinuously *over* time. In particular, scholars and activists have drawn attention to Indigenous groups whose lands have been systematically colonized, and to whom apocalypse—in the sense of the destruction of their livelihoods and cultures—means something much more immediate.[55] Many Indigenous cultures have already experienced the epochal shift of ecological destruction and genocide associated with normative notions of apocalypse in the Anthropocene; the idea of extinction as a purely theoretical future event, therefore, becomes a fallacy.

Yet the ways in which the Anthropocene is made globally legible—through news reporting, scientific study, and cultural artifacts such as film and fiction—are overwhelmingly skewed toward a Western orientation. This has been a common cultural trope seen through a proliferation of "cli-fi" books, artworks, and big-budget movies depicting a ravaged human populace in thrall to a vengeful planet.[56] Some scholars have pointed out that this particular narrative might help us face the enormity of our destructive actions while simultaneously allowing us to generate future worlds not dependent on inequality and destruction.[57] Although there is merit in this viewpoint, what it misses are the discontinuities in both experiencing apocalyptic effects of environmental destruction and engaging with how the apocalyptic might be different for different actors.[58] By constructing apocalypse as a future event that will wipe out most planetary life-forms, this popular configuration of doomsday is rooted in Judeo-Christian eschatological notions of redemption, alongside technoscientific fetishization that cultivates the fantasy that even if the planet starts acting up, a white man will undoubtedly come along and save us.[59]

Permafrost becomes subsumed into this narrative as a particularly insidious harbinger of doom, cropping up more and more frequently as stranger and stranger things start happening. The Pleistocene Park works with this global vision of permafrost through its rallying cry to "save the world" for humanity, while remaining keenly aware of its more localized quirks as the Zimovs battle constantly to keep their project afloat—literally, given how frequently the Kolyma floods as a result of warmer Arctic winters. In "Earth," I critically engage with the idea of human extinction through an interrogation of permafrost as a global object that foreshadows a future planetary apocalypse and note the ways in which this narrative falls apart upon closer inspection. Human extinction is not the same as planetary apocalypse, but the two become conflated in attempts to mitigate permafrost thaw for the benefit of human survival, as in the Pleistocene Park project. The permafrost has been dubbed a globally significant "ticking time bomb," but this misses the localized yet still catastrophic outcomes of permafrost thaw: collapsed houses, flooded towns, anthrax outbreaks. A concept as nebulous as human extinction necessitates a more nuanced understanding of the planetary category of "human." Locating the experience of apocalypse in the permafrost landscape of Sakha and demonstrating how the town of Chersky can be conceptualized as an island reveals that the threat of extinction is missing a number of multiscalar, multitemporal processes that are not necessarily defined by Western concepts.

Arctic Extinction

There is a sense of hubris to thinking about human extinction in the Anthropocene, even if the goal is to decolonize and dismantle the structures of privilege that saturate apocalyptic imaginaries. Indeed, the idea that humans might someday go extinct seems to come with the expectation that the planet and all its life-forms will also cease to be. This is, of course, highly unlikely. As this is a book largely concerned with permafrost and its becoming discontinuous in the Anthropocene, a fundamental question must be: Can permafrost go extinct? The question can be further broken down once the initial resistance to such an alien concept has been overcome: What does it mean for an inhuman substance to "die," and what are the other forms of loss that emerge from an absence of permafrost? Two different modes of inquiry are necessary to begin to understand whether permafrost can go extinct, and they depend on different understandings of what permafrost *is*—specifically, for the purposes of the question at least, the multiple materialities of permafrost and the multiple ontologies of permafrost. These epistemological interventions are not necessarily at odds, but they do highlight the different ways of knowing permafrost, and how they will always refuse to form a whole—permafrost is, ultimately, discontinuous in many forms.

Permafrost suggests permanence. Disregarding the active surface layer that thaws seasonally, permafrost is largely expected to stay put. That is why the Soviet Union plowed so much money and effort into studying permafrost, and why cities like Yakutsk were able to be built, with stilts sunken into the ground to support buildings so their heat does not melt the soil beneath, pipes built aboveground, and so on.[60] That permafrost is now thawing, retreating, slumping, and buckling was not in the plans of the Soviet scientists and engineers who built those roads and towns; that the permafrost might cease to be—if not soon, in the distant future—has implications not only for the people who make their homes on top of it but also for the material landscape of Sakha itself. Radin and Kowal point out that "freezing is simultaneously an act of creation and an act of destruction"; surely the material inverse is also true.[61] Thawing both creates and destroys strata, manipulating the landscape into simultaneous vistas of novelty and decay. Permafrost is not an inert state but rather encompasses intersecting multiplicities that are shaped by the dynamic and discontinuous processes of thaw.[62]

The liminal material state wrought by permafrost thaw and its concomitant processes of being "in between" materialities—through slush, mire, bog, or slump—is testament to a slippery form of agency that is clearly not radically separate from the forms of life that are found within the permafrost landscape.[63] In discussing her concept of "geopower," Elizabeth Povinelli points out how the divisions between life and nonlife are fueled by human anxieties around the precarity of survival, which is increasingly dependent on a subtending geology.[64] Similarly, Kathryn Yusoff argues in her work on what she terms "anthropogenesis" that the origin story of the human is inextricably geological, and that ecological destruction reveals the tension at the heart of survival in the Anthropocene. She states: "What is at stake is not a nature that involves entities *per se*, but what passes between them, holding together or forcing apart."[65] Extinction in this sense cannot be defined by the assumed division between life and nonlife; the permafrost thaw that threatens earthly survival is not a separate entity but rather a process that curates the trajectory of extinction in all its forms.

What does this mean in the context of the Anthropocene? Permafrost occupies a materiality that is different from that of rock, which encloses fossilized remains for posterity. Thawing, conversely, reveals and removes. Take the example of the thermokarst megaslump, the largest of which is the Batagaika crater in Sakha: a hole that is growing consistently larger, revealing a transient stratigraphy that renders visible the region's climate and ecological history, if only fleetingly. Sequestered carbon filters into the atmosphere; mammoth carcasses begin to rot as some are scooped up by hunters seeking the precious ivory of the tusks; ancient viruses escape. When thinking of an Anthropocene legacy, it is necessary to consider not merely what is added to the earth but also what is taken away. Mark Carey categorizes glaciers as "an endangered species," purposely muddying the waters of what counts as an extinction as a way to highlight the narrative of loss and "saving" that accompanies climate change discourse in the Anthropocene.[66] If extinction can be defined as the end of something, thinking through the retreat of matter brings about a material shift, or perhaps absence, of a landscape that was once there but is no longer. The permafrost's becoming discontinuous is a rupture of planetary significance that forces introspection into the relations between human and planet, and blasts through the anthropocentric and normative definition of extinction. It also demands an interrogation into the resurrective designs of the Pleistocene Park—what does it mean to "restore" an ecosystem that nobody can remember?[67]

It is not just a material absence that suggests a permafrost extinction but also the loss of particular ways of life and understandings of what permafrost *is*. Indigenous groups, largely nomadic, made their homes on the tundra long before the Soviets began moving in, and the Sakhans migrated from the Baikal region to the Lena delta area in the south of Sakha, where they began horse breeding and other forms of agriculture that worked with the permafrost landscape. The Sakhans and the various smaller Indigenous groups—Even, Chukchi, and Yukaghir mostly—practice a shamanic cosmology, in which spirits play a fundamental role in how the world is ordered and lived with. In Sakhan mythology, the world is split into three realms: the underworld, the surface, and the heavens. These realms must be finely balanced in order for Sakhans to live harmoniously with their environment. Permafrost acts as a barrier between the underworld and the surface (where humans live), the integrity of which must be treated with respect, otherwise evil spirits can escape. Julie Cruikshank's ethnography of Tlingit and Athapaskan glacial oral histories characterizes glaciers as brimming with agency—mischievous, resentful, perceptive, and dynamic.[68] While Western narratives of glacial environments have romanticized their pristine and awesome wilderness, to Indigenous groups glaciers are social spaces where humans and landscape mingle. The sentience of glaciers as described in oral histories points to ways of engaging with the landscape as an agential and relational force that must be balanced with human activity. Any attempt to disregard such a balance may be met with destruction—for both human and glacier. The same is true with permafrost.

That permafrost is known almost exclusively through Western scientific methods speaks to the need to decolonize how certain knowledges are privileged over others and demonstrates how making space for other forms of knowledge can help open up the definition of extinction.[69] But this is not to say that inhuman extinction cannot be experienced and understood through Western ontological positions. In recent years funerals have been held for several glaciers that are now considered to be extinct, with participants including a variety of actors, such as local people, artists, and activists. Such practices represent divergent forms of loss, but the losses are not necessarily comparable to species extinction.[70] The point is that permafrost thaw and environmental destruction are experienced in multiple ways and often through attachments to specific landscapes or places—in this case, Sakha. Thinking about extinction as more than merely the death of a species necessitates inhabiting a discontinuous space that recognizes different and often fractured ontological understandings of what permafrost is. The thawing of permafrost not only has material effects, but it also produces new ways of understanding the world—and, by that logic, destroys old ones. In "Ice," I tell several stories of permafrost thaw and the multiple configurations of permafrost life that emerge from this fracturing landscape. By paying attention to the spaces left behind by thaw, we might begin to find forms of extinction that are surprising, unsettling, but also generative and creative of new ways of knowing permafrost, revealing that inhuman forces cannot be extracted from the lives that make the permafrost their home, nor can the politics and power differentials be set apart.

Nonhuman Extinction

Species extinction—that is, the extinction of species that are not human—is the form of extinction we are most familiar with and, some might argue, used to. There is no way of knowing the actual numbers of species extinction, given that no one can say definitively how many species have existed in the world; estimates of the rate of species extinction per year range from 0.01 percent to 0.1 percent—much, much higher than the average background rate.[71] With the acknowledgment that the sixth great extinction is due to human activity, it is vital to note that the pressures on species life and extinction are underscored by processes of globalization, unfettered capitalism, and habitat destruction, contextualized by unequal power relations. The sixth great extinction demands an interrogation into not only which human practices (and who wields them) are responsible for species death but also which species are valued over others when it comes to conservation—or, indeed, commodity value.[72] Jennifer Telesca's work on the bluefin tuna encapsulates the tension at the heart of extinction in the Anthropocene: because of competing systems of value, the life of the bluefin is in thrall to an extractive system promoted by capitalist institutions.[73] Through an understanding of the political and economic relations behind the ambiguity of the sixth great extinction (which works in a homogenizing way similar to that of the Anthropocene), species death is revealed to be a much more specific and targeted practice. While it may seem counterintuitive, it is these same capitalist systems of value that curate the polar bear clinging to a diminishing ice floe as the representative of the climate crisis.

The specificity of species death also matters in how extinction is responded to. Just as the responsibility for the sixth great extinction cannot be meted out equally across all humans, the notion of some sort of collective grief, or eco-anxiety, is a fallacy that has taken root in Western environmental discourse; this feeling of helplessness in the face of global climate change and extinction tends to erase specific histories of settler colonialism in favor of a vague future-oriented catastrophe, not to mention the loss of animal lives and histories that are attached to more localized places.[74] To pause for a moment and think of what it actually means to lose a species forever requires exploring beyond the categorization of extinction on a red list or a statistic and thinking through the more gradual and less conspicuous degradation of relations and kinship built throughout generations.[75] Thom van Dooren advocates for the telling of stories that both speak to the mourning inherent in extinction and convey the liveliness of the creature that might be captured through the retelling of its "becoming with" other entities.[76]

Understanding extinction as a slow process of undoing time makes it more difficult to pinpoint an exact moment of loss; indeed, extinction is already much more slippery than that. Cary Wolfe suggests that extinction can have already occurred even if the last members of a species still remain—material reminders that extinction is a process of *unbecoming*.[77] Dolly Jørgensen's work on the concept of the "endling"—the last individual of a species—highlights the painful experience of witnessing not only the death of an animal but also the rendering obsolete of its entire history, including its entangled relations with places and people. The endling—the most famous of which is the last passenger pigeon, Martha, who died in 1914 as the final representative of what had been one of the most abundant species on Earth—takes on a reverberatory power that forces humans to confront the enormity of extinction as a historically contingent process.[78] Similarly, extinction can be viewed in a more local context, understanding that attachment to place and processes of extirpation (losing a species in a particular geographical area although it still exists elsewhere) might produce a sense of loss that manifests itself as a sort of haunting within the landscape.[79] The presence of the physical, breathing body of a creature might be how existence is defined in conservation terms, but the remains of a species or a group of individuals—whether these be material remains in the form of bones or more spectral remains in the form of place-names, departed nests and dens, memories, and stories—reverberate long after the species has been classified as extinct.

The Pleistocene Park is a confluence of these difficult, fuzzier extinctions, drawn out across millennia, and which might still be considered to be ongoing. That the Zimovs felt compelled to restore the mammoth steppe suggests that they believed something was missing from the landscape, and this was further compounded by the permafrost beginning to thaw. The discovery that the Pleistocene megafauna were essential to maintaining the integrity of permafrost while they simultaneously stimulated a rich grassland with their bodies reveals the long arm of extinction; nobody alive, of course, can remember the Pleistocene ecosystem, but it makes itself known through the absences of the creatures that produced it and the bones they left behind.[80] This type of restoration ecology, or rewilding, which focuses on restoring entire ecosystems rather than conserving endangered species, has grown in popularity in the past two decades. While its advocates argue that it can foster an enchantment with wild landscapes that are increasingly being lost, its detractors—with whom I share some concerns—are uncomfortable with the subconscious message that landscapes free of humans and human activity are somehow better than those that are not.[81] Such an idea upholds the homogenizing undertone of the sixth great extinction, which lays the blame at the feet of all humans and harks back to a romanticized past. The Zimovs do not claim to be promoting some mystical wild ideal, but their designs on restoring the mammoth steppe rest on two assumptions: that humans should be absent and that the animals they bring to the park perform an ecological function devoid of any attachment to place or history.

What happens when these animals return? What happens when bison trample the bones of their long-dead "ancestors," or a newly de-extincted mammoth unearths a buried tusk? The Zimovs have already rewilded several species that have been extirpated from the region for decades or longer—some brought from downriver, some transported by truck from thousands of miles away. If extinction is a slow unraveling, then does a rewilded animal begin to knit the frayed edges back together, or has the absence in between produced something else entirely? The process of restoration here is a discontinuous one that attempts to rejoin generations of kinship amid the ruins of a lost ecosystem. Whether the park is successful in its mission remains to be seen, but the world is a very different place in the Anthropocene than it was in the Pleistocene. Ben Garlick's work on the historic displacement—and subsequent ostensibly successful return—of ospreys in Scotland points to the "lived spatio-temporal particulars of osprey life," which are not easily replicated by mere rewilding.[82] While the example of ospreys speaks to a particularly culturally geographic landscape, to pluck a herd of baby bison from Denmark and expect them not only to become Siberian but also to take on the role of forgotten epochal "relatives" ignores the reverberations of extinction that extend across space and time.

The rewilded animals at the Pleistocene Park are tasked with "saving the world." This world is the world of humans, and the task is to prevent human demise. Nonhuman extinction is often curated around the needs and narratives of humans; even attempting to tell the stories of extinct and endangered creatures, as extinction scholars advocate, can easily become self-indulgent.[83] In "Bone," I focus on the skeletal histories of the various bones found and revealed by permafrost retreat and how they can act as material registers of extinction within both deep and shallow timelines. The bones themselves have afterlives. While they are relics of a lost ecosystem, preserved for millennia by permafrost, their revealed presence produces echoes that vibrate across timescales and create both a sense of anxiety and an imagined utopic past. What the Zimovs are trying to do with their project is to reproduce this fantasy utopia as a response to the growing fear of a dystopic future. Their roughshod frontierist approach to science is indicative of a heteropatriarchal dynamic that is predicated on legacies—both their own and those of the animals they curate. What the permafrost does, however, is trouble these linear legacies through processes of material and temporal discontinuity; I will argue that this ruination offers a chance for confrontation and, perhaps, an opportunity for something new to emerge from the ruins.

De-extinction

The final form of extinction I address in this book is one that exists mostly as speculation, although this is rapidly changing. The idea of de-extinction has existed as a trope in fictional media for some time, the most famous example being, of course, Michael Crichton's *Jurassic Park* (both the 1990 novel and the film based on it), in which things go horribly wrong on an island full of resurrected dinosaurs meant as a tourist attraction. It is no coincidence that the Pleistocene Park draws more than a little inspiration from de-extinction fiction, although undoubtedly Sergey could not have ever imagined the possibility of mammoths wandering his Pleistocene tundra when he christened the park in 1996. Mammoths tend to be touted as the "poster animal" for de-extinction, being arguably the most charismatic and, indeed, the most dead.[84] Most candidates for de-extinction tend to be species that have recently gone extinct and for which large caches of genetic material are available from bodies or museum exhibits; mammoths, in contrast, are a viable option because the preservational properties of permafrost have made them so. Currently several laboratories are actively working on mammoth de-extinction through different methods. George Church's laboratory at Harvard University is working on a form of hybridization by using the gene editing technology CRISPR to plug mammoth genes into an Indian elephant genome. Church has recently spearheaded a new de-extinction organization dedicated solely to mammoth de-extinction; called Colossal, it is bankrolled to the tune of $15 million by the CEO of PayPal and claims, as of this writing, that it can resurrect the mammoth within five years.[85] Meanwhile, scientists at Sooam Biotech in Korea and Kindai University in Japan believe that there is a piece of mammoth DNA to be found in permafrost that is so well preserved a mammoth can be cloned.[86]

Regardless of which method is used, the idea of de-extinction surely completely upends the basic tenet of extinction: that it is forever. Resurrecting an extinct creature, alongside other scientific advances in the life studies such as gene editing and cloning, demands a reevaluation of ethics and responsibility regarding the conservation of endangered creatures.[87] Two main points of tension arise from the possible use of de-extinction as a valid method of conservation: one argument is that important resources and effort will be diverted from attempts to conserve at-risk creatures if the option exists to just "bring them back"; the other takes the moral standpoint that resurrection biology is something that humans simply should not meddle with—the "playing God" argument.[88] This standpoint opens up a whole new raft of concerns: Who gets to decide which animals are de-extincted and when? Why, exactly, are some animals chosen over others? De-extinction becomes seen as a reductive practice, in which life is merely something that is, or is not, granted. These are questions of value, in which it must be asked whether any resurrected mammoths would possess a life on *their* terms rather than on the terms of their so-called creators.[89] In a sense, de-extinction encompasses the paradox of the Anthropocene: that the epoch of humans designates (some) people power over life, yet the circumstances of the sixth great extinction come with the caveat that humanity is unable—or refuses—to create a "safe operating space" for planetary survival.[90]

De-extinction falls firmly within the remit of a so-called good Anthropocene—a call for the titular Anthropos to act as environmental stewards, benevolently steering Earth processes into harmony through science and technology.[91] Geoengineering is also indicative of a good Anthropocene, as is rewilding, to an extent; the Pleistocene Park is a proponent of all three. The environmentalist Stewart Brand's now-famous quip, "We are as gods, so we have to get good at it," encapsulates the attitude toward a good Anthropocene that puts its faith in science but fails to take into account the structural power differentials, white supremacy, and unending capitalist thirst that underpin the glorification of a Western techno-utopia.[92] It will come as no surprise that Brand is heavily involved in de-extinction, and his organization the Long Now Foundation, which advocates for long-term technological projects, has a branch called Revive and Restore that generously funds George Church's work. In their sprawling 2012 book on synthetic biology *Regenesis*, Church and Ed Regis state: "The mammoth almost cries out for resurrection. Some specimens unearthed from permafrost are so lifelike that they appear to be merely sleeping, not dead, much less extinct."[93] De-extinction arguably represents the pinnacle of a good Anthropocene, in that it offers the ability to conquer death and extinction and to conjure life at will. Who gets to be the magician, however, will invariably align with the vested interests of the powerful.

De-extinction forces a confrontation with what life is, where it is found, and what it means to create it. Extinction, at its most basic level, is essentially the end of a species through the mass death of its population. De-extinction, conversely, works at the level of the gene, whether that be through CRISPR hybridization or cloning, both of which work with viable genetic material that has been preserved, most often by freezing. The cryogenic freezing of cells is essential to de-extinction, particularly for the mammoth, whose "natural" permafrost freezer has begun to fail; mammoth remains must be cryogenically frozen in order to maintain the integrity of the DNA. This raises the question: Has extinction even occurred if the cellular material endures across time and death? Adam Searle identifies this paradox as "the liminal ambiguity of de/extinction, at once the archetypal case of reversing extinction, while at the same time the clone of an animal from anabiotic cells that had never been extinct."[94] Life becomes molecularized, located at the level of the cell; extinction now becomes a genetic issue, not a social one—something to be paused, turned off, or, in some cases, even turned on.[95] As a response to the sixth great extinction, the cryobank becomes a machine through which life and death are recategorized and commodified, time is suspended and remade, and immortality becomes—ostensibly anyway—a possibility.

Of course, the fundamental question of de-extinction is whether any so-called resurrected creature could be deemed to live the "true" life of its kind: the mere physical presence of an elephantid creature with shaggy fur and cold-adapted blood does not a mammoth make. A de-extincted mammoth would be removed from any temporal and geographic relation to her extinct ancestry; most likely a hybrid, she would be born to a mother of an entirely different species—another hugely controversial and unethical practice—and almost immediately plucked from her care to be placed in a strange land, expected to fulfill an ecological function she has no knowledge of. She would initially also be the only mammoth in existence, occupying the role of an almost reverse "endling," yet remaining stuck in a liminal state as a representative for a not-quite-extinct creature: a ghostly figure that haunts both the mammoth's past and its future.[96] Deborah Bird Rose speaks of a "deathzone" as a multispecies space of encounter in which the reverberations of extinction must be confronted.[97] What confrontation will de-extinction generate, particularly when the mammoth's resurrection is buttressed by the weight of expectation that she will reprise her role on the permafrost?

In "Blood," I address the fantasy of immortality that de-extinction provides, but within a human context rather than a nonhuman one. De-extinction is a human science after all, albeit one curated through modes of power and capital, and the mammoth is afforded resurrection only because of her ability to create livable conditions for humans. The notion of living forever goes hand in hand with recategorizing humans—as Stewart Brand does—as gods. This way of thinking is indicative of an Anthropocene that underscores the dominance of certain humans yet ironically exposes the creeping dread that this dominance—and its accompanying exploitative way of life—might be coming to an end. If de-extinction is used as a way of maintaining a level of mastery over planetary processes—or, indeed, "saving the world"—it becomes imperative to question just what world is being referred to here, and if it necessarily should be saved.

Applying Discontinuity to Extinction

There is a growing need to engage critically with the Arctic in a way that resists its "climate barometer" status, through its specific cryopolitics of coldness and the heterogeneity of its geography. Tim Ingold calls for a "Northern ontology" as a way to move beyond the Arctic's status as a region, to practice a definition of "Northness" that is not applicable at other geographic coordinates but is instead its own milieu of specific materialities, naturecultures, histories, and futures.[98] It aims to take ice, and its concomitant properties of freezing and melting, as a precondition of Northern/Arctic being. Sverker Sörlin speaks to the intersecting temporalities of ice; while its melting is indicative of a short-lived tension, ice also has the capacity to affect longer climatic time frames. Sörlin states: "In this sense we have reached a 'cryo-historical' moment—*cryo* signifying ice and snow, directing our attention to the historical powers of human forcing in the Anthropocene."[99] A discontinuous approach aims to take these ideas as a starting point to capture at least some of the difference, multiplicity, and surprise that roots permafrost in place—that place being the tundra and permafrost landscape of Northern Sakha. To consider how a situatedness within non-Western ontologies produces different ways of recounting and telling stories of permafrost, I take inspiration from anticolonial scholars who are either Indigenous themselves or work closely with Sakhan Indigenous cosmologies.[100] Western imaginaries of Arctic and tundra landscapes point largely to their emptiness and isolation, depicting them as places of sterile wilderness save for the odd intrepid male explorer or scientist; the scientific singularity of climate change and Anthropocene scholarship has painted glacial and permafrost environments as giant laboratories in which to conduct experiments that support key geological theories while maintaining a sense of frozen sterility.[101] By contrast, the Siberian Arctic is a place of dwelling and Indigenous lifeworlds.[102] It is a place of animal life, albeit a depleted or rewilded one. It is a place of bloody history and a place of uncertain futures. It is a place of discontinuous permafrost, tundra, mammoth steppe—many words for the same, but also different, thing.

Thinking specifically about a permafrost materiality is similarly necessary when applying discontinuity, which is why I have named each of my chapters for a material component of a permafrost landscape. Underscored by new materialist scholarship and engagement with the agential properties of the inhuman, thinking discontinuously about matter pushes back on attempts to control or categorize permafrost.[103] Permafrost in the Anthropocene, particularly that close to the surface, is an almost constantly shifting state change, a confounding substance that refuses to be permanent. Permafrost is both a chunk of frozen soil I can hold in my hand and a vast tundra landscape stretching across the Northern Hemisphere.

This encapsulates the fundamental problem of the Anthropocene's grand narrative: how to speak to both a planetary story and a heterogeneous, fractured, local one. By drawing on permafrost stories that encompass different voices, perspectives, ontological worldings, knowledges, and scales, writing discontinuity produces a catalog of collaborations that emerge in different ways. They are reproduced here in a purposely jagged way, not necessarily in a linear format, as a way to create a sense of almost disorientation that cleaves to the unpredictability of permafrost. These might take the form of retellings, anecdotes, occurrences or events, or a slow decay across time; they might be grand narratives or smaller happenings; they could be official documents or oral histories. Storying is not merely a representational practice; it is also a material one that gathers together or holds apart various worldly orderings and relations.[104] What storying can also do is draw attention to multiple permafrost ontologies and ways of living on, and with, permafrost that are not scalable or reducible to a singular experience. Russia, the Russian Arctic, and Sakha all possess specific, if interlinked, histories that press upon the present and future of the permafrost in different ways. Not only this, but the permafrost *produces* the untimely by refusing to fall into anthropocentric temporalities. Its undulating freeze/thaw materiality, at once ephemeral and ancient, theorizes the permafrost as something that breaks free of human categorizations. A discontinuous permafrost materiality underscores these stories, preserving histories (both recent and not) and casting futures. It is all the researcher can do to attempt to listen and, perhaps, to retell them.

Permafrost in the Anthropocene is oriented toward a reductive and specific mode of survival, straddling a boundary of apocalyptic catastrophizing, wherein the "strange" things the planet seems to be doing are harbingers of human extinction, and a sort of doubling-down effect, in which a Western consumer culture refuses to relinquish the spoils of global capitalism. Discontinuous thinking disrupts the existential dread that is coded into the Anthropocene's designation of geological agency. By making interventions into time, scale, and knowledge, discontinuity aims to shatter modernist assumptions of linearity and boundedness. When permafrost is mobilized as catastrophic and rendered as a global object and planetary condition, the failure of science to understand it fully—and thus maintain control—curates a looming future apocalypse that demands the world be saved.[105] To think discontinuously disrupts the progression of geology *toward* the future, notices the retreat and decay of matter across different scales, acknowledges the unknown, and recognizes different temporalities. The singular, future apocalyptic event brought on by an unruly permafrost mass does not exist when thought through discontinuously.

But discontinuity is not merely a refusal of things; it is also generative. Discontinuous interventions into matter and ontology highlight permafrost's dynamism and creativity, offering inspiration into how worlds that resist the normative, largely Western construction of the planet are found within a deeper understanding of permafrost. By acknowledging the liminality of permafrost, with its cycles of freezing and thawing, or its unpredictable retreat, thinking discontinuously resists the assumptions that matter is controllable and predictable and that nonlife is in direct opposition to life; in committing to becoming *non*committal, we might begin to unravel extinction in its heterogeneous, creative, and ungraspable forms. Extinction is not a singular, linear event, nor does it occupy a single definition. In this book, I aim to offer a reorientation away from normative understandings of life, death, and extinction by drawing attention to instances when these understandings fall apart. Taking inspiration from the permafrost's refusal to provide the ideal conditions for life, I will advocate for discarding the continuous birthing of power structures and the quest to control life in the Anthropocene and instead embracing the surprise, the multiplicity, and ultimately the discontinuity of our discontinuous Earth.

1 Earth

On a swelteringly hot day in the Russian Arctic, I accompany Sergey and Nikita to their field site on the tundra behind the science station. We are hurled around as the Land Rover traverses buckled and degraded permafrost roads, at points almost fully reclaimed by encroaching thermokarst lakes.[1] Sergey explains to me how these lakes increase ground temperature by acting as an insulating blanket. "Like an iron," he says, simulating the steam with a hissing sound. Eventually the road disappears completely and we abandon the car to continue on foot, Nikita shouldering a heavy motor and drill while I carry the permafrost thaw depth probes. We are here to test the Zimovs' suspicion that the permafrost in this area, normally a solid layer that never thaws, is showing signs of sporadic but long-term degradation. Arctic tundra permafrost in northeastern Russia is uniformly categorized as "continuous," whereas permafrost found in the more boreal landscapes of the southern taiga can be patchy and therefore "discontinuous." To find discontinuous permafrost this far north is an unprecedented—and terrifying—prospect. Sergey, an old man and a heavy smoker, is struggling in the heat; periodically, he gathers a handful of snow that has endured under rocks and washes his face with it. Sometimes, he tells me, he strips off his clothes and places hunks of earth on his body, making a sort of cooling blanket from permafrost. He is an Arctic man, not made for these uncharacteristic temperatures. As we stumble forward through clouds of mosquitoes, Nikita points out abandoned structures, clear-cut areas, overgrown roads to nowhere, even a forgotten runway. "There used to be all sorts of industry here," he tells me: hugely productive mines and factories, and even a military operation to transport ballistic missiles. "It's good to know that if you really wanted to you could just bulldoze the ground and go 'fuck the permafrost.' " He is only half joking.

The first field site is an open stretch of nonforested tundra, the ground covered with scrubby mosses and stunted shrubs. Without shade or tree cover, it is a likely spot for increased ground temperature and permafrost thaw. As the sun beats down, our feet sink slightly into the mushy ground. Sergey attaches a gigantic drill bit to the motor and starts to burrow down. The drill slides through the active layer easily, like butter, and then starts to shudder as it hits the harder soil beneath. He leans bodily on the machine, forcing it downward as far as it will go, and then withdraws the drill. The silver drill bit emerges from the earth caked in mud. Into the hole goes the thaw depth probe, Nikita applying his whole weight to the thin metal pole. The difference between the length of the drill shaft and the depth achieved by the probe is measured with a tape, and the results are scribbled in Sergey's notebook with a pencil. The Zimovs do three of these drill-probe tests and then we move on to the next site. We spend the baking hot afternoon moving from tundra to larch forest to *alaas* and then back to the car.

Sergey and Nikita check the permafrost thaw depth probe to see the extent of the annual thaw on the tundra.

The two men agree: there is definitely something up with the permafrost this year. The conversation at dinner that evening centers on how best to communicate these findings to the wider science world. While the initial readings offer scant evidence of an overall trend, Sergey is adamant that this is no anomaly; he has noticed the temperatures increasing for several years now, and he has to warn everyone. The message is clear: what the permafrost is doing now is unprecedented, and if left unchecked, permafrost thaw will spiral out of control. The work done by the Pleistocene Park now takes on greater significance, compounded by a lack of both money and time. Sergey is keen for an article about the thaw to be published in *Nature* or, at the very least, *Science*. Reporters for *National Geographic* are going to be visiting in a few weeks to do a feature, and they will be told all about it. And then there is the matter of the TED Audacious Project competition, the winner of which could be offered as much as $100 million to fund an "audacious" idea that tackles a pressing global issue—an audacious idea like turning the whole of Siberia into an Arctic savanna.

We film the video appeal for the TED grant a few days later. Nastya, Nikita's wife, is behind the camera, while Sergey and Nikita sport matching Pleistocene Park fleeces and media-ready grins. We do multiple takes, some ruined by a gust of wind, the ringing shot of a hunter in the distance, a passing car, an overzealous dog wandering onto the "set"; some are rendered useless by the Zimovs stumbling over their lines. They follow a script of stripped-down, easy-to-understand language; the video has to wow the judges in under a minute, so this is no time to mince words. The tone is deliberate: "We see only one way to stop this global climate catastrophe—we must bring back [the] ecosystem of mammoth steppe." Not only does this video need to appear exciting, but it also has to communicate to a largely indifferent global audience that permafrost science is an essential part of climate change research, and what the Zimovs are doing is more than just interesting, it is necessary. The very basic and almost childish language used in the video betrays the fact that many people are ignorant about what permafrost even is, let alone that it is in trouble.

Permafrost, as a substance that covers 25 percent of the Northern

Nastya films Nikita and Sergey for their entry in the TED Audacious Project competition, which, if successful, could net the Pleistocene Park one million dollars.

Hemisphere, is of global concern—especially now, as its thawing will have global consequences. But watching Nikita and Sergey struggle to find the right tone to convey such a sentiment in a place so isolated that locals refer to the rest of Russia as "the mainland" is a jarring experience. How to square this dimpled bit of headland, and its scraggly, spindly trees and moss carpet, with the fact that permafrost as an area of land stretches from the diamond mines of Chukotka to the forests of Alaska? How to reconcile the thaw depth probe sinking deep into the earth with the frightening rate of global climate change? These are questions of scale, and they are intimately entangled with discussions and knowledge-making practices of the Anthropocene. That the Anthropocene concept tends to globalize and homogenize both the planet and the human species is now a well-worn critique from scholars across multiple disciplines, and this is precisely what happens to permafrost when brought into view as an Anthropocene signifier.[2] Permafrost maps, almost ironically, tend to be more "global" than standard maps of the world: seen from a top-down viewpoint, the North Pole is centered with the spherical relief of landmass outlines curved round it, permafrost shaded in some icy color such as purple or blue—a world away from the dense, almost tacky mud I can scoop out of the ground in Chersky and roll into a ball with my fingers.

These scalar renderings have implications for permafrost knowledge too. How permafrost is studied, imagined, and produced is dependent on the situation or context. Permafrost experts of all scientific persuasions consider permafrost as a substance differently than a journalist writing for popular science media does; a person living in the Russian Arctic perceives permafrost—or tundra—differently than a person living in a warmer climate does. There is a disparity across these knowledge-making practices, and the understandings that emerge from them enter into different arenas or spheres of influence. The Anthropocene is not formed solely in the halls of the Royal Geological Society; rather, it is enacted in multiple ways and on multiple scales, from different vantage points and ontological positions.[3] Permafrost knowledge bleeds across these boundaries. Permafrost exists as a continuous mass on a map, but that is only one of the myriad forms it can take; emerging from fraught histories of Soviet oppression and colonization, it travels fifteen time zones caked on the boots of an American ecologist and buckles the airport runway in Chersky. These different manifestations of what permafrost is, and how they become funneled or discarded as part of a wider narrative of catastrophic climate change, are underpinned by power and politics that are mobilized on a global scale to generate anxieties regarding human survival. In particular, they form part of a longer trajectory of apocalyptic dystopia that has been growing in the environmental movement for decades, coalescing not at a single point in time, like an asteroid, but rather as a series of events that have built in severity *over* time.[4] As the threat of climate change looms larger, so too does permafrost become curated as a global apocalyptic object—a danger to human life, an "x-risk"—rendered through particularly Western forms of knowledge making. Though smaller instances of permafrost thaw (no less devastating to the lives affected) are largely discarded in favor of a grander narrative, they can nonetheless generate ruptures in the Anthropocene's trajectory:

discontinuities that redefine apocalypse and extinction in a way that challenges assumptions of linear and homogeneous life and time.

A Global Permafrost Apocalypse

The double meaning of this chapter's title is deliberate. *Earth* and *earth,* the same word but encompassing vastly different scales. The various practices of sensing and mapping the planet have historically produced a territorial understanding of the Earth, parceling its landmass and, often, its ocean spaces into various forms of politically sanctioned ownership.[5] Recently, scholars have problematized the way territory is produced as obscuring the volatile terrain of a planet composed of heterogeneous and shifting materialities—particularly in the wake of the climate crisis. Stuart Elden revised his influential work on territory to account for the importance of terrain, or, specifically, the "nonhuman agency of the earth."[6] The climate crisis is often portrayed as a planetwide phenomenon, but the specificity of matter and how it both generates and is affected by climate is key to recognizing how global heating is unequally distributed. As Gastón Gordillo puts it: "Global warming reminds us that materialism teaches us one simple truth, confirmed by our mortality: that matter always-already exists in excess of our human perception and uses of it."[7]

That does not mean that human methods of categorizing matter do not attempt to account for more volatility. The past decades have seen a proliferation of organizations and task forces using climate models and scenarios to predict trends and attempt to manage the risks of climate change; these efforts seek to categorize the planet as predictive, conforming, and, most of all, under our control.[8] Global summits are held to form agreements on thresholds and limits, put strategies in place to affect quantifiable levels of carbon dioxide, and produce temperature graphs. Past climate events, each bounded by a beginning and an end, are used to generate predictions, a process culminating in linear plots of possible future scenarios. Permafrost, however, is not included in most large-scale climate models; it represents too much of an unknown to be useful.[9] Its seasonally shifting materiality, unpredictable and sometimes dramatic slumps, sensitivity to "abnormal" weather events such as heavy snowfall and forest fires, and heterogeneity across landscapes mean it is very difficult to categorize. As one permafrost scientist stated in 2018: "The models can't handle those landscape-scale changes, all of the processes that could lead to rapid change, and it's going to be a long time before they can."[10] Similarly, attempts at mapping permafrost point to the transience of the data produced: any map created is almost immediately out of date.[11] It is the uncertain space found in this liminality that constructs permafrost as apocalypse: "The logic of apocalypse is marked by a certain anti-epistemology—the impossibility of knowing."[12]

The actual likelihood of a tipping point that results in the complete loss of permafrost across the planet is unlikely, however. Speak to any permafrost scientist, and they will be hesitant to sensationalize in the way that popular media articles do. One of the United Kingdom's top permafrost scientists, Professor Julian Murton, almost winces when I broach the subject of tipping points during a visit:

> We know that permafrost can be preserved through climate conditions as warm or warmer than present over timescales of hundreds of thousands of years, so at least that gives some reassurance if you're a permafrost person that permafrost isn't going to be disappearing or whatever, even if the climate gets five degrees warmer than present; thick, cold permafrost is still going to be around in ten to twenty thousand years' time. The surface of it may have thawed a bit, creating various structural engineering problems and releasing carbon and things, but the very thick, several-hundred-meter-thick permafrost, is still going to be around. There's not going to be a sudden "one year" that has like little effect, and the next year like there's going to be a major effect. So, if there is a tipping point it's quite a slow tipping point. . . . Even if there was really rapid warming in permafrost, it would thaw—the top of it would start to thaw, but it's not going to happen all like that [snaps fingers], it's going to be a slow process.

This is not to say that permafrost scientists are looking to play down the severity of the situation. Permafrost thaw is already causing many problems for daily life in Sakha, particularly through damage to infrastructure. Permafrost scientists are very worried indeed about the trends exhibited by permafrost in both the continuous and discontinuous zones, although they also point to the lack of data available, particularly in Russia.[13] The point is that every prediction made, every model run, also exhibits a huge unknown.

Much of this has to do with the heterogeneity of permafrost as a materiality: more a patchwork of different and difficult material processes than a sweeping overlay on a world map. Permafrost slumps, creeps, karsts, shears, heaves; it forms polygons, contains ice wedges, produces lakes and rich *alaas* ecosystems.[14] The things frosted ground can do are numerous: blister, boil, burst, churn, crack, fissure, pry, shatter, shift, split, and suck. Despite the important work done by the scientists who visit NESS and the long-term studies of Russian permafrost researchers, what permafrost *does* when it thaws and where it does it is difficult to track. While much of this has to do with the vast—and often isolated—spatial reaches of Arctic permafrost, at the heart of it, permafrost can be surprising. It is not easy to predict when and where slumps will occur, and in the case of the Batagaika crater, the name that has been given to the rapidly widening hole, Doorway to the Underworld, is representative of the sort of legend curation that tends to emerge when things are unexplainable.

So how does this square with the apocalyptic visions promoted by the Zimovs and, in turn, by popular media reporting on the goings-on in the Arctic? The global media tend to report on permafrost through the language of urgency, making reference to tipping points and "ticking time bombs," and to hidden things awakening from the ice; such reporting often supports more generalized discussions of climate change or Arctic warming.[15] The popular media coverage speaks to a shadowy future event that contains a kernel of potential annihilation. A certain eschatological bent is found in the way the weird and forbidding things emerging from permafrost are reported, and a subterranean, often patchy substance made volatile by climate heating conjures up an imaginative explanation for any physical manifestation of permafrost thaw. In the imaginations of humans, the subterranean has long housed all manner of demonic entities; a cursory glance at the news today can give the impression that they have finally escaped.[16]

What globalizing the apocalypse does is ignore the very intimate and often local entanglements of place and specific materialities and instead produce a singular relation between human (the species, but also the individual savior) and planet. It subsumes the specificity of permafrost and its liminal freeze/thaw and slow durations. It envelops the creeping aridness of previously fertile land. It disregards anything that falls beyond the remit of a homogeneous human experience: "Certain types of humans (capitalist, imperialist, hyperconsuming, enslaving) . . . have started to imagine our own end . . . as the end of the only thing that we deem to be worthy of the notion of 'world': us," states Claire Colebrook.[17] It constructs an end, a big bang, a mass death of technocapitalist life as we know it. Ends are important here.[18] What the Anthropocene concept also does is construct geological time within the boundaries of beginning and end; by designating epochal status, the Anthropocene joins the ranks of the Holocene and the Pleistocene as an enclosed time period. Yet how and when the Anthropocene will end remains unknown, with an underlying assumption being it can *only* end with the accompanying end of the "Anthropos."[19] Extinction and apocalypse, as Thomas Moynihan reminds us, are not the same, but by funneling climate disasters toward a catastrophic end for humanity—and by this, I am occupying a loose definition of "end" that encompasses societal collapse, not necessarily full human extinction—we are able to kick the can down the road.[20] Imagining a future apocalypse caused by permafrost thaw obscures the very real and current problems being experienced by many who dwell within the permafrost landscape, or orients them toward Western concerns. There is a reason the Pleistocene Park declares there is still time to "save the world": admitting that the Arctic is melting faster than we can dam the hole unsettles the idea that it is possible to shepherd the planet; constructing an apocalypse that has not yet happened permits the fantasy that there is still time to stop it, to wrest back control from an unruly Earth.

Immersing myself in life at the science station and in the town of Chersky, I gradually become aware of how the narrative of catastrophic permafrost thaw and global apocalypse emerges from the tundra soils of this isolated corner of Northern Sakha. The practice of doing science on the ground is different from what makes it onto the page and into internationally renowned journals. The patchwork of intersecting and often less recognizable practices can become hidden in the language of permafrost science, or in the keen eye of a cameraperson looking for a dramatic shot. These stories—the ones that do not make it into the grand, sweeping vistas of the documentaries or the careful scientific studies dealing with quantitative results—are what I came for; to understand just how permafrost has come to be bound up in this particularly "Anthropocenic" apocalyptic drive, and how the homogeneity of global thinking obscures the current and ongoing crises of permafrost thaw around Chersky, we need to go to ground.

Permafrost Science at the Pleistocene Park

I am the only visitor at the science station in the first week, and the ice is still melting in the Kolyma, thick slabs drifting slowly down from the Arctic Ocean. There is a sense of slowness, of awakening, to that first week, before the trickle of international researchers and media crews begins in earnest. The park itself is inaccessible until the ice has fully melted, and many of the sites that scientists use to monitor permafrost and collect data are cut off for weeks on end; timing one's very expensive visit to the station is therefore key. I arrive with an intention of responding to permafrost from the ground up; rather than attempting to slot permafrost into my schedule, I will adapt my work to the rhythm of temperature. This is much to the confusion of the Zimovs, who are used to receiving structured timetables from their guests, but as the weeks of my stay pass by, I realize my methods and the methods of the visiting scientists are not so different after all. They, too, are dependent on favorable climatic conditions and the availability of equipment; we all rely on permafrost making allowances for our presence. Nastya remarks to me one day that her favorite time of year is when the ice is just beginning to break up on the river. "It sounds like singing," she says. During that first week, I spend a lot of time watching ice.

After a week, two things happen: the ice melts to the point that the town—including the airport—floods, and the first scientist arrives at the station by emergency helicopter. He is an ecologist who studies Arctic tundra vegetation, and this is his tenth consecutive field season. Others have been coming to the station every summer for even longer. They are scientists from all different disciplines—geography, ecology, hydrology, organic chemistry, ornithology—but their work has one thing in common: permafrost. Even if they are not here to study permafrost specifically, their work necessarily engages with permafrost by way of their doing fieldwork and engaging with the landscape. Scientists who spend a field season at a research station like NESS are very much embedded into the permafrost landscape; living on top of the tundra informs both their work and their day-to-day lives. If the ice is late to break up, they will be prevented from reaching their field sites. If a forest fire occurs, they will have to change how they approach their measurements. If their instruments stop working over the winter, they will not be able to collect complete sets of data.

Most scientists who visit NESS are working on long-term studies that track changes to an isolated parameter over time and space. The helicopter ecologist, Dr. Mike Loranty, monitors Arctic plants in areas of high-density and low-density tree cover. Dr. Heather Alexander shows up a few weeks later with a group of students in tow for her fifteenth year at NESS studying the effects of forest fires on permafrost soils. A German contingent funded generously by the Max Planck Institute has sealed off an area of the park to track carbon emissions from soils. Ornithologist Dr. Eugene Potapov returns year after year, heading out for weeks at a time to the thermokarst lake–dotted Kolyma basin to monitor bird migration patterns. These scientists come from all over the world, brought together by Arctic permafrost as a research framework. They bring with them expensive and specialized equipment, much of which is left behind in the soil laboratories at the station. The work they do is rigorous and disciplined, methodologically designed to pass through peer review and into top science journals; their knowledge is very much legitimated by their being part of an international scientific community.[21]

American permafrost scientist Mike Loranty in the field with his "little bit of kit." He has visited NESS every year for ten years to conduct a long-term experiment on permafrost vegetation.

And while at the dinner table they might confess their worry about climate change and permafrost thaw, the language of their science is restrained and muted.

I accompany Mike to his field site, which has been set up to document how vegetation disturbance due to climate change can alter permafrost thermal dynamics.[22] His instruments dot the tundra, contained in plastic bags to protect them from the winter freeze; his backpack is full of spare batteries to replace the ones that have run down since last summer. He has also brought a drone with him to make maps of the area, and he spends his days painstakingly flying it over the terrain in neat little grids. What becomes apparent as I observe several of his forays to the tundra to change batteries is how reliant he is on circumstance and luck. We wait in silent tension as he boots up the first data logger, which has been monitoring vegetation all winter; anything could have happened to make it stop recording, and he tells me of past bitter disappointments—meltwater seeping into batteries, locals coming along and stealing his equipment. There is palpable relief when it turns out the data logger has endured, and I cannot help but be struck by how dependent Mike's science is on local conditions. "Sometimes I feel amateurish working here," he says. "Have you been to the German site, did you float by there on the way to the park? It requires so much more money and it's just so much more sophisticated. . . . I feel kind of amateurish with my little bit of kit."

Heather arrives at the station a couple of weeks later with more than a "little bit of kit," a team of research assistants, and a huge budget. She and Mike are old friends, as might be expected given their sustained involvement in the same patch of isolated Arctic tundra. But despite the large scale of her project, she remains reliant on Sergey's intimate knowledge of the local area to pick a field site where her team can monitor forest fires and soil burn severity.[23] There has been a fire relatively recently just down the Kolyma, and the group will take the station's barge there to collect samples. Forest fires caused by lightning strikes are a common occurrence in the Arctic, but they have becoming increasingly frequent with climate change and are having a knock-on effect on the permafrost.[24] While the work of both Mike and Heather has global implications and international appeal, the locality of their efforts and the specific outcomes of research into this *particular* permafrost landscape mean there is a disconnect that does not allow for a scaling up of knowledge.[25] The permafrost soils and ecosystem of Chersky are different from those of Alaska and Canada. When permafrost is considered as a global object of global significance, it cannot be separated from the careful, painstaking, lengthy research that takes place at the local level.

The various methods and approaches to doing permafrost science converge around a particular point—in this case, the science station and the park. There is an almost awkward coming together in the way we all sit around the dinner table, the conversation centered on permafrost in a rather patchwork way made up of divergent knowledges.[26] We speak anecdotally, share rumors and gossip, and extrapolate how climate change will affect permafrost next year, in ten years, in fifty. We help each other: Mike advises Sergey on how to structure his paper for *Nature*'s rigorous peer review, while I proofread Nikita's application form for the TED Audacious Project competition. The reality of living and working with permafrost may produce a sort of discontinuity as we all bring our piecemeal knowledges to the table, but permafrost as a knowledge *object* is rendered scientifically legitimate only through specific and *singular* scientific practices and instruments—for this knowledge to go out into the world, the more ad hoc methodologies must be discarded.[27] Isabelle Stengers points out the homogenizing process a particular thing must undergo to become known as science: "Science, when taken in the singular and with a big *S*, may indeed be described as a general conquest bent on translating everything that exists into objective, rational knowledge."[28] Permafrost science, being done in a collaborative—and, indeed, discontinuous—way at NESS must shed any notion of multiplicity and knowledges not centered on scientific parameters if it is to go out into the world as a singular knowledge object. What happens to this knowledge as it moves from a local space of heterogeneity into the Western scientific arena is a scaling up—a process of globalization that amplifies and legitimates data and trends while simultaneously obscuring the nuance of local permafrost knowledges. The park acts as a sort of crucible for these competing knowledge scales, encompassing the specificity of climatic minutiae but also presenting a highly globalized vision of apocalypse and planetary redemption.

The Pleistocene Park as Planetary Redemption

The science station is how the Zimovs make their money, and it is an invaluable base for international scientists conducting Siberian permafrost research, but the Pleistocene Park is what grabs the attention of the global media. Access to the park can be difficult: the trip from the station takes thirty minutes in the boat, and that depends on someone being available who can take you, not to mention the environmental conditions have to be just right. I am kept waiting a week before Nikita judges the ice on the Kolyma to have melted sufficiently for him to make his first visit to the park this season. When we arrive, the park looks nothing like the one seen in documentary films and glossy science magazines, the kind found in dentists' waiting rooms; it is brown, scrubby, mostly flooded, and there are no animals to be seen except for a blind yak and a couple of reindeer. The journalists and filmmakers who make the trip here are tasked with presenting the potential of an Arctic plain replete with prehistoric-era creatures, and they employ all manner of camera and drone trickery to find this: sweeping shots of tundra, slow-motion close-ups of animals, even computer renderings of mammoth herds on the move.[29] Yet their narrative is equally a story of broader, human proportions, centered on the almost David-and-Goliath motif of two men challenging the might of the Russian wilderness. Their footage emphasizes the pioneering nature of the Zimovs as they face the encroaching risks of climate change and the fragility of permafrost—a highly gendered portrayal that pits the masculine Russian hero against a disobedient Earth.[30] The permafrost as an unknown and unpredictable substance, making its presence known through worrying and unexplainable events, is tied into a visual narrative designed to reach people who have no scientific knowledge themselves. It is not enough that there is a rewilding project somewhere in Siberia; the rewilding project must *matter* to people. The rewilding project must do the work of saving the world.

Cody and Kenneth show up at the station as part of Heather's group. They are student filmmakers, here to follow the scientists into the tundra and film them working in the field, conducting experiments and taking samples of permafrost. Their final product is to be a documentary aimed at high schoolers, an attempt to make climate change matter to a potentially apathetic audience. The film's trailer juxtaposes shots taken on the trip with recent news clips of climate-related incidents—Hurricane Sandy, California wildfires, Florida flooding. I ask the filmmakers why, and Cody responds:

> We chose to do that simply to create connection. Siberia is a foreign place with very little tangible connection for someone in the United States. But climate change is happening in all of our own backyards. . . . The series itself is about how this research connects back to what people already know, but in a different scope.

> Science is boring to most people, and many people don't trust them or relate well to scientists.

Cody's opinion that people find science "boring" is apparently shared by those responsible for the apocalyptic fearmongering found in most popular science narratives about the permafrost and climate change in general. There is no doubt that permafrost is thawing at a greater rate than is normal, but in order to convey the severity of the situation to audiences that have never seen permafrost, reporters resort to hyperbole while minimizing the difficult scientific language that largely cannot offer any certainty. This is exactly what Cody and Kenneth are trying to do, and why the film crews that come to NESS focus on grand, sweeping narratives. No longer painted merely as a substance found in an isolated region somewhere up in the North where very little happens, permafrost is now imbued with apocalyptic possibility.[31]

This globalizing of the Zimovs' project taps into both the unpredictable nature of permafrost and the encroaching dread of apocalypse. It is entrenched in the promotion the Zimovs do, both at international conferences and online through crowdfunding sites such as Kickstarter and Indiegogo. In order to offset the almost prohibitive cost of transporting animals to such an isolated region, the Zimovs have run crowdfunding campaigns with slogans such as "The World's Best Plan to Bring Back a Vanished Ice Age Ecosystem and Save the World from a Catastrophic Global Warming Feedback Loop" and "Bison to Save the World."[32] The slogan of the park itself, printed on memorabilia such as mugs, T-shirts, and shot glasses, is "The Proper Tool to Fight Global Warming." Although the park currently occupies only a small area of northeastern Siberia, the Zimovs' ultimate goal is to expand to such an extent that the mammoth steppe would range across the entirety of Northern Russia. In essence, the globalizing of the park's motivation underpins the expansive vision of the project but also the urgency of global warming; by framing their appeals through global concern, the Zimovs are curating a sense of the planetary that uses the threat of apocalypse as a way to encompass all of humanity and its potential extinction. For the Pleistocene Park to access the necessary funding to take it from a local ecological experiment to a rewilding project that can affect planetary processes, they need to scale up their story. And while Nikita is keen to downplay his heroic mantle in conversation with me, he does not shy away from the world-building element of his work:

> You know, I have three kids; they will be growing up in this world. I cannot make an island here. I cannot. They will be part of this social community and they will be facing all the challenges of the world. In fifty years, I'm probably dead, and by that time they're on their own and their kids are on their own, their grandkids . . . I'm really worried about the whole world. Many people ask: "Do you really think you can make any difference?" Their main concern is that can you really make ecosystem big enough to have climate impact or anything, and many people don't believe that. I am not sure we will succeed with what we're doing. I don't know what will be in the future, I don't have any guarantee. I cannot guarantee you that we will have mammoth steppe saving the planet in the future, but I can guarantee you if I will not be doing that, that mammoth steppe will *not* save the future—that's for sure! I know that the road is very long and complicated but the only way to follow this path is to just start walking that direction. It's better to walk rather than to sit and wait for death.

Nikita's allusion to the "islandness" of Chersky chimes with how the locals view their town as cut off from the "mainland," but what Nikita is also doing is enlarging his project *and* his situatedness onto a construction of the planetary that is legitimated by the scientific work done at the station. This is very much in line with a scientific singularity that constructs the globe from a distinctly Western perspective, and one that has been forcefully assimilated across the world through the brutality of colonialism and empire. The Pleistocene Park's globalizing narrative of planetary redemption homogenizes *both* the permafrost and humans as a species, stuck fast to a runaway climate catastrophe.[33] What is happening here is the construction of an apocalyptic climate event (albeit a slow one)—a homogenizing force that categorizes the human species as an amorphous collective facing the rage of the planet together, culminating in a dramatic explosion that might spell the end for humans as a species. The Zimovs see their role and the role of the park as defusing the ticking time bomb of permafrost, but escaping the boundaries of their island and scaling up their project is not as easy as their crowdfunding campaigns suggest.

Chersky and the Mainland

The more time I spend at the station and around Chersky, the more I become aware of my situatedness, and the more attuned to the slightest changes in temperature, temperament, and the crossing of various paths. Nikita knows that to achieve his vision the park must break free of its island confines, but at the moment, that is not possible. The material rhythms of this island life are dictated by barriers (albeit permeable ones), by being cut off from the rest of the world. The knowledges and narratives produced here at NESS bleed outward through the work of scientific singularity and popular media, but on the ground they remain, for now, contained by permafrost. The islandness of Chersky is anchored not only to the town's history of brutality and hardship but also to the very ground on which it was built. Its soils, once rigid enough to support a population of almost twelve thousand, are now sagging under the weight of crumbling and abandoned buildings; roads are potholed and saturated with meltwater, with power lines lurching to the side. On a particularly hot day, I watch the twice-weekly plane take off shakily from the earthen runway, creating a cloud of dust so thick the twin-prop jet is soon swallowed up by it. The runway used to be asphalt, but in recent years the permafrost thaw has become so bad, the poorly funded local government has not been able to afford to make repairs whenever the runway buckled. While the long winter allows for the use of ice roads on the frozen Kolyma, in the summer months, the plane is the only way into Chersky, which is bordered on one side by roadless tundra and on the other by the Arctic Ocean—an island if there ever was one.

Named for the Russian explorer Ivan Dementievich Chersky, who met his death in the area in the late nineteenth century, the town is one of the many seaports dotting the Russian Arctic coastline that were key to the Soviet Union's plan to "master the North" through economic and cultural colonization.[34] Most of the residents are thus ethnic Russians, while Sakhans are more likely to be found farther south. Initially, the port of Chersky (then known as Nizhniye Kresty) was built in the 1930s to transport gold down the Kolyma from the nearby mines populated by gulag labor. Today the town's small museum, housed in the former high school's gymnasium and open by appointment only, carefully tracks Chersky's difficult first century through a large and well-curated collection of artifacts. Unlike many museums in Russia, it does not ignore the atrocities of the gulag, and one corner is dedicated to the people who lost their lives under atrocious conditions of forced labor and servitude. The exhibit includes references to "Bloody Lake"—the site of a mass murder committed by gulag officials in response to an attempted prisoner uprising in 1937. The prisoners spared were not permitted to bury the dead, whose bodies reportedly stained the waters of the lake red; after several days, they were interred in an unmarked mass grave. The soils of Chersky are more than just permafrost; the earth is more than just earth.

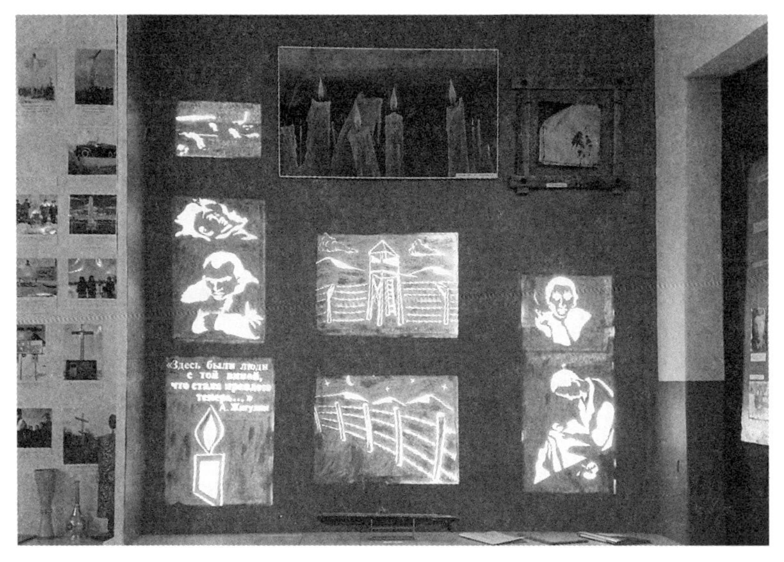

The local museum, housed in the gymnasium of the closed-down high school and open by appointment only, holds a wealth of material on the history of Chersky, including a section dedicated to the victims of the local gulag.

After perestroika, Chersky's economy collapsed. The jobs, the infrastructure, and the state-subsidized perks, such as helicopter shuttles and cheap gasoline, dried up.[35] Within ten years of the Soviet Union's dissolution, the town had lost two-thirds of its population to out-migration.[36] Now, many of the buildings are empty and jobs are scarce; food prices have exploded because of the high cost of transit. The people who stay are those who can afford to live here, or those who cannot afford to leave—ironically, the islandness of Chersky contains an uneasy tension between those who profited from the collapse of the Soviet Union and those who lost everything.[37] The Zimovs were among the lucky ones. While Nikita recalls the 1990s being very tough on the family, he tells me proudly that NESS was the only Arctic research station in Russia that managed to keep operating after perestroika. Whereas most had relied on the subsidized helicopters to transport researchers to their tundra field sites, Sergey had had the foresight to invest in diesel boats. Diesel was not exactly easy to get ahold of back then, but it was a lot easier to come by than helicopters.

After many years of economic hardship, the Chersky residents are now facing a new problem. The permafrost earth, the very foundation of the town, is slowly becoming its destroyer. This is a pattern seen across the Arctic, but in Russia in particular, the damage caused by thawing permafrost does not seem to be a top priority for the government. My visit coincides with the reelection of Vladimir Putin to the presidency for a fourth term, and I am told Chersky has fielded the highest percentage of votes in Sakha in favor of the incumbent. I find this difficult to understand, given Putin's apparent lack of interest in the plight of increasingly cutoff Arctic towns, but Nastya provides the answer: "They know that if they vote someone else, he could be even worse. At least this way they know what they are getting." Nikita merely says, "In Russia, democracy is a burden."

Chersky's islandness is constricting and its attachment to the mainland increasingly threadbare, yet, thanks to the Zimovs and the Pleistocene Park, the town is progressively interconnected. The visits of international scientists and global media have put Chersky on the map, but in a way that presents it as a warning for the rest of planet rather than as a specific locality bound by a bloody past and a fraught present-future. The permafrost becomes a disembodied entity, stripped of its layers of intersecting histories, its underpinnings of power and politics discarded in favor of a singular narrative: that the permafrost is thawing, and human earthly survival could be at risk because of it.

Human survival is already at risk. Chersky residents, many of whom rely on the land for their livelihoods, are finding island life increasingly precarious. The Indigenous people who live in Chersky and congregate in the town's cultural center know of enforced precarity all too well—the nomadic reindeer herding they had practiced for centuries was systematically broken down by the Soviet state, collectivized and assimilated in the name of cultural unity. The history of Russian colonization across Siberia can be found in the arrangements of people and towns throughout Sakha and the Arctic, and the ways in which certain livelihoods have been altered or swept aside. In a study published in 2011, a Nenets man is quoted regarding his reaction to the "newcomer" to his lands:

> He really, probably, is happier than me. After all, he does not worry, does not root for the devastated forest, for the forest dug by a bulldozer, for the river littered with the road. . . . And after him all my life I have to live here on dug, trampled earth, with broken trees and with oil-polluted lakes and rivers. I will have nowhere to get furs and meat, nowhere to catch fish for slicing, nowhere to graze deer. Isn't this an apocalypse?![38]

The narrative of apocalypse in the Anthropocene similarly homogenizes the effects of climate change on the planet, focusing on some vague future catastrophe and discarding the multiplicity of lived experience. Elizabeth DeLoughrey, writing specifically on islands, argues that we must "provincialize" the Anthropocene as a way to account for different views that refuse a colonized world map.[39] Indeed, islands have come to be seen as powerful "holding sites" of the relationality and friction of Anthropocene living, as ways to challenge both global and linear thinking.[40] Islands provide nodes of rupture or indeterminacy to the perpetuation of a homogeneous worldview, with their ecologies and cultures acting to anchor their specific locations and complex histories.[41] Denying these specificities, as the Zimovs do in their quest for funding, produces a singularity of permafrost that misses its discontinuity—and the discontinuity of the planet—as part of a localized milieu. The islandness of Chersky might exhibit wider trends of permafrost thaw, but the reverberations felt by the people who live there are anchored stories of apocalypse in their own right.

Permafrost Ruptures and Island Apocalypses

When the airport in Chersky floods, I hear the alarm from inside the basement of the cultural center and feel that familiar, yet unfamiliar, creeping dread aroused by the sound of a wartime air-raid siren. All flights are canceled until further notice, until the water levels drop and the runway dries out; helicopters are drafted to ferry people in and out (at their own expense) if they desperately need to travel. The Kolyma normally floods with ice meltwater every summer around this time, but this year the river reaches the highest levels anyone can remember. In other parts of Sakha, people post videos to social media showing their houses underwater, people being airlifted to safety, and sodden dogs clinging to rafts. When we finally make it out to the park on the boat, we are met with a veritable tide of destruction: ranger huts are partially submerged, scientific equipment has been washed away, and park animals have disappeared into the larch forest to escape their waterlogged feeding grounds, so Nikita has to go and find them. This flooding is the result of unprecedented snowfall during the previous winter that blanketed and insulated the earth and has now burst the banks of the river. One might be tempted to think that the ground's inability to absorb this meltwater means that the surface soil has a robust permafrost underbelly as impermeable as solid rock. But actually, heavy flooding increases the ground temperature and encourages thermokarst, generating a feedback loop of escaping greenhouse gases and increased air temperatures. It is likely there will be another heavy snowfall next winter.

These occurrences are ongoing and often unnoticed by the world beyond Chersky, flaring brightly and then fizzling away again with the coming and going of the seasons. Even Moscow is selectively

The Kolyma River bursts its banks a few days after I arrive, flooding the Chersky airport and other lowland areas and cutting the town off from the "mainland." The flooding is due to the heavy snowfall of the previous winter, which was particularly warm.

blind to what is happening, preferring to delegate provincial issues to cash-strapped federal governors. The locals deal with whatever comes their way with a sort of stoicism, but only because there is little else they can do. I watch Nikita's face as he is told that his permafrost tunnel, complete with several seasons' worth of permafrost cores and other data, has filled with water for the first time ever. He rolls his eyes and gives a short laugh. "We are fighting global warming," he says philosophically, "but global warming is fighting back." His reaction is indicative of how life is lived on the Siberian tundra: there is no point in getting angry in response to the shifting and often surprising whims of the climate; you just have to deal with it. Without a large-scale qualitative study of attitudes toward global warming, it is difficult to know how the residents of Chersky view these climate events, but Nastya tells me of a recent regional survey in which residents were asked whether they believed the permafrost was thawing because of climate change; the results were a landslide

Leonid, a fisherman, lives alone on the banks of a Kolyma tributary and welcomes visitors who bring food and vodka in exchange for the folk songs he plays on his guitar. His fishing lake has flooded, but he does not believe in climate change.

of climate change denial. A couple of weeks later, on the summer solstice, we visit Leonid the fisherman on the way to Duvanny Yar. He breeds hardy tundra dogs and serenades us with Russian folk songs, accompanying himself on a guitar Nikita brought back for him from San Francisco. Leonid's fishing lake has breached due to the flooding and he has lost all the fish—his livelihood—but he pours us shot after shot of vodka and reassures us that things will all be back to normal next season. He does not believe in climate change; to him, "it's all just weather."

There is clearly a disconnect between the vast, global narrative of climate change hurtling the planet toward its doom and the smaller ways the Chersky locals respond and adapt to their changing environment, living in a very particular ecological and climatic condition that an Anthropocene planetary focus misses. This is northeastern Siberian Arctic permafrost, and while the permafrost as a mass object has global implications for climate change, the entanglements that arise from the shifting materiality of frozen ground take place at a localized level. It is the very discontinuity of permafrost behavior that generates these localized responses; a thermokarst slump here will be quite different from one a few hundred miles to the south. The unpredictability of permafrost's behavior from year to year and the seasonal fluctuations that follow no clear pattern or prediction in a turbulent climate make permafrost living a complex art. The ways the locals react to these fluctuations also follow no clear pattern, but they are certainly not oriented toward a buildup of certainty or prediction; they are not gathered for use in a climate model or pointed to as signs of a coming global apocalypse. How to view these quieter, less "important" reactions to permafrost thaw? How do they square with a homogeneous and bounded apocalyptic event? Through a consideration of these discontinuities as a sort of rupture in the fabric of the steady, modernist vision of the world, the smaller and perhaps obscured knowledges of difficult permafrost living are afforded space.

The notion of rupture has traditionally encompassed large, world-shattering revolutions, and as such has become weighted by the history of failed revolutionary projects—and, staying in place for a moment, the Soviet Union is an obvious example of this.[42] Yuri Slezkine writes in *The House of Government* that Bolshevism was merely another millenarian cult in the vein of Christianity, Islam, and many other utopic sects.[43] The utter annihilation of what came before Bolshevism was the one path to Marxism-Leninism and was supposed to culminate in a glorious day of reckoning in which utopia would materialize on Earth. The ultimately failed project of the Soviet Union was a patchwork of generative intersections, each propelling or stalling the end goal of redemption, culminating in the annihilation of what came before—in a sense, an apocalypse that generated both renewal and catastrophe. Much of this rested—and still rests—on the material landscape. Much more than any other nation, Russia encompasses permafrost into its built environment—partly for economic reasons, but partly because there is so much permafrost, it simply has to—while the tensions of the Cold War rendered permafrost as a military arena, a battle space of both temperature and territory.[44] The Soviet Union ultimately imploded, but its legacies reverberate through the current ecological anxieties of the Anthropocene.

Chersky is undoubtedly one of the many places in the Russian Arctic where the shallower ruptures of the Soviet Union's dissolution and the longer, slower decay of the material landscape converge. This presents an opportunity to recast rupture as less of an event and more of a gradual encroachment of disruption, or, as Martin Holbraad and his colleagues put it: "The unsettling effects of rupture, understood as a radical and often forceful form of discontinuity, are central to current perceptions of a world in turmoil."[45] The way to introduce rupture back into the theoretical fold so that it recalls its historical baggage but also breaks free from the notion of the bounded, linear political event is to pay attention to what the plurality of the word can do. The events I have just recounted are the smaller discontinuities, the less revolutionary ruptures, the ongoing and entangled apocalypses of banal register.[46] The Soviet project was indeed a resounding failure, but its dying throes sent reverberations all the way to the town of Chersky, as evidenced by the town's 1990s mafia gangs, the decay of its seaport, the out-migration of almost 80 percent of its residents. Climate change and permafrost thaw produce material effects that rupture ground and buildings but also the idea of a singular, continuous permafrost. The threat of global apocalypse emerges from the thawing permafrost soils of Chersky, but they also contain the remnants of a thousand other, smaller, overlapping apocalypses that produce the town's specific island locality.

Reimagining the Planetary

The potential for human extinction is necessarily tied to globality. On the one hand, this underscores humanity's pervasive dominance in the Anthropocene, but on the other, it reveals a certain sense of entrapment—an involuntary commitment to this insignificant third rock from the sun. As the *human* epoch, the Anthropocene, and its accompanying fear of apocalypse, curates a sense of the planet that, in opposition to the expansive utopic dreaming generated by the early space exploration of the 1960s, becomes uncomfortably inward-looking. Claire Colebrook pulls no punches when she states: "It is the image of the globe that lies at the center of an anthropocentric imaginary that is intrinsically suicidal."[47] By remaining stubbornly wedded to the globe as the quantifiable unit of sustaining life, the growing sense of dread that the planet is becoming increasingly resistant to human survival means that we tend to imagine it less as spaceship Earth and more as a floating tomb. Notwithstanding Elon Musk's designs on "colonizing Mars," the notion that we are all stuck fast on a planet hurtling toward its doom cements not only the rootedness of earthly living but also the inevitability of earthly dying.[48] The figure of the tomb—that which is underneath, while the majority of human existence takes place on the surface—conflates with the precarity of thawing permafrost. The exposure of the subterranean through thaw is a process of disruption that lays bare the vulnerability of planetary living. The globe as an *idea*—that pale blue dot looked back on from space—must be thoroughly disrupted if we are to hack down into the depth, the volume, the underground necessity of earthly life.

To imagine a globe is to imagine a spherical surface—not quite flat, but rather the shiny curve of schoolroom prop smooth to the touch. Similarly, to recall the maps of permafrost is necessarily to engage with a sense of shallowness, all the paper or pixels can allow. Just like the knowledges and the voices that are obscured in climate science, the more localized and smaller differentiations of permafrost become subsumed into this flatter, homogeneous mass on a rendered globe. But permafrost has depth; it extends, takes up space, has volume. The artist Nikolay Smirnov, in writing about an exhibition on permafrost that he curated in 2016, uses the term *planetarity* instead of *globality* as a way to better reconcile the local with the global through a "complex material surface."[49] A concept initially introduced in the work of Gayatri Chakravorty Spivak, planetarity renders a world that is beyond human control to shake free of the homogenizing forces of globalization and capitalism. Spivak simply states: "I propose the planet to overwrite the globe."[50] Overwriting does not replace but rather creates layers, reveals differences, makes space for islands, and draws out disparities between the hegemony and those it obscures. Planetarity makes space for the subterranean and does away with the smooth topography of globes; it welcomes the multiplicity of a permafrost materiality that changes seasonally, exhibits surprising and dynamic features, produces effects that are experienced differently locally yet still have implications globally. We might call it earth, or a discontinuous Earth.

A thought experiment: let us invite an alien archaeologist to examine what might be found in permafrost a century from now, an extraplanetary visitor with the scanning technology to examine depth and volume in incredible detail. A piece of dried fish from Leonid's hut, the carcass of a bison that did not survive the winter, a rusting ship hull—all find their way somehow into perpetuity in permafrost depths to be discovered and analyzed as markers of the Anthropocene.[51] These are found artifacts, but there are other registers too, of a fractious ontological positioning: the burrow of an underground mammoth god, a subterranean world of evil spirits as described in the Sakhan epic *Olonkho,* the ruins of an experimental ecosystem.[52] These fragments do not necessarily form layers, as rock does, but rather sink or emerge with the churn of permafrost and what that dynamic force does to assumptions of space and scale. Permafrost can be deep but also shallow, its thaw dependent on a variety of factors, not all of them human-caused. Paying attention to multiplicity of scale offers an escape from the definitions of large and small, global and local, or even vertical and horizontal. What permafrost does is produce a disruption in what it might mean to consider both the discovery of a long-discarded thaw depth probe and the redrawing of permafrost features on a map. Our alien archaeologist is faced with both a top-down view from her spaceship and an earthier, smaller excavation on—and in—the ground.

Disrupting a linear time frame by imagining an archaeologist looking *back* at Earth—indeed, a planet not her own and of a completely different species and ontological positioning—upends the notion of a vertical fossil record of stacked strata. Current Western thought places the temporality of the Anthropocene as bookended by its beginning and end, made legible, as are those of all other geological epochs, by the fossil record, strata upon strata, the traces of which reveal a potential extinction event coded into the rock. An epoch as a measurement of geological time is necessarily human in origin;

the fundamental fear produced by the paradox of the Anthropocene is that this is the final epoch. The planet may continue on without humans, but epochal time will not. Any coming global apocalypse closes the future; it destroys not only humanity but also the ways humans have found to quantify temporal life on Earth. When Nikita speaks of "waiting for death" if he does not strive forward with the Pleistocene Park, he is constructing the future as a coming tide of destruction that crashes back on the Newtonian arrow path of time, sweeping away humanity and the world as we know it. It is this unknowable "other" that produces the future imaginary of human extinction.

The notion of a global permafrost object compounds the idea that all permafrost, everywhere, is thawing. But permafrost is not a homogeneous object; our future archaeologist can see how bits have remained for millennia, as a patchwork of peaks and troughs. If the Anthropocene encodes human existence within its rock record, permafrost strata are more transient, and much less discriminatory when it comes to preservation. As the Batagaika crater grows ever larger, the animals in the Pleistocene Park trample their little fenced-in area and the permafrost beneath endures. Elsewhere in Russia, anthrax spores from a long-dead reindeer escape into the atmosphere and kill a child.[53] Down the Kolyma, a patch of wildfire-scorched tundra begins showing signs of regrowth.[54] The difficulty of permafrost research and the specific language used highlights how scientists are aware of their unknowns; to think in terms of a permafrost tipping point is to imagine permafrost as a singular material, responding and reacting in consistent, predictable ways, alongside the fantasy that something can be done to reverse it. It suggests a sort of linearity, with the substance sliding down a temporal slope in one direction. What would happen if we applied discontinuous thinking here? What if we imagined, instead of a tipping point, feedback loops—*temporal* loops—of smaller, messier permafrost materialities, all happening at different times, at different durations? These incomplete temporal loops unsettle the idea of the Western configuration of time as quantitative, as minutes stacked upon hours stacked upon eons.

This has implications for theorizing apocalypse. The bounded event, with beginning and end, is unsettled by these ruptures that can flare up quickly, leach out slowly, or occur unnoticed, deep underground. They draw together disparate temporalities; they reveal histories and erase futures, as they also might erase histories and reveal futures. The freezing and thawing of permafrost is a product of both cyclical seasonal time and incomplete temporal looping; the friction generated by these temporalities might be termed an apocalypse, albeit one that is difficult to define. Abandoning the homogenizing configurations of the "Anthropos" and its conflated potential extinction event means that different forms of apocalypse spring up. When the planetary is reoriented geographically and historically to account for the Soviet creep eastward and northward, other timelines and cartographies emerge. Apocalypse, after all, means an "uncovering," a revelation of knowledge. Why not consider the widening expanse of the Batagaika crater as an apocalypse? Why not the burning bodies of anthrax-riddled reindeer? The flooding of Leonid's fishing lake? These smaller instances, permeable events with no defined beginnings or endings, disrupt the continuity of daily, monthly, yearly rhythms. They create surprise, which is constantly being responded to, adjusted to. They do not rain fire from the sky or flood the entire planet, but they affect, and are affected by, the discontinuous materiality of a shifting permafrost earth.

The Lives All Around

A couple of weeks into my stay, the container arrives. The huge shipping container, packed to the brim with supplies by Nikita, left Novosibirsk around a month before for its long, slow trip up the Kolyma to Chersky. Because of various environmental factors—weather, ice melt, currents—the barge carrying the container has been late arriving, so we have been surviving on hunted elk meat and produce from the prohibitively expensive local shops. Most of the container's contents are for the visitors to the science station, but I notice that there are quite a few other shipping containers as well as outbuildings around the complex, and I question Sergey about what they contain. "Oats," he says. "For horses?" I press. "For humans," he replies. Another morning at breakfast, as I try to piece together what the Zimovs believe is happening to the planet, Nikita tells me he is ready at a moment's notice to decamp the entire family from their winter home in Novosibirsk to Chersky for safety. "We are waiting for that black day," he tells me without further explanation. For all his refusal to make an island of the park, Nikita knows that Chersky's isolation would be a benefit in a doomsday scenario he clearly believes is coming. On the evening that Mike arrives, we stay up late after dinner, and Sergey confesses his fear: "I can't sleep at night. I've so much worry about the whole of civilization." He has noticed how the past few winters have seen temperatures reach above zero; they have never done that before. He reveals he is stashing away food and other supplies, as well as selling his assets for gold; he talks of his sheds full of supplies and the abundance of fish in the Kolyma, repeating a traditional Native saying that all a man needs for survival is a deerskin sleeping bag, a deerskin coat, and a small shotgun. One can imagine that Sergey keeps a bug-out bag of these items under his bed.

The international scientific community's response to the Zimovs' findings detailed at the beginning of this chapter has walked a fine line between skeptical and worried.[55] Many scientists are concerned about the study's lack of methodological rigor, with the only tests conducted so far consisting of the insertion of a thaw depth probe in a few locations over a single summer season. They point to the absence of long-term temperature data and analysis of historical trends. These are the scientists who track specific changes over long periods before forming any conclusions, and those conclusions are likely to come with a "but." Conversely, Sergey has the courage of his convictions, and that makes him a difficult man to work with.[56] Nikita explains: "My dad is a very provocative scientist. He's always trying to say things the most aggressive and offensive way. . . . As you can guess it's extremely hard to pass through the peer review." Nikita's stance is to serve as a mediator between the force of nature that is his father and the wider scientific community, bound by rules, methods, and peer review. He sees the value of scientists returning to NESS year after year to take measurements, and not only for the much-needed funding they provide; he uses every opportunity he can to repurpose instruments and nudge ongoing experiments toward his own research interests. But as much as he feels frustrated at his father's almost bullish nature, Nikita has learned everything he knows from Sergey, and he believes in his father's vision wholeheartedly. "He's by far the most brilliant scientist I know," he admits. At the time of this writing, the Zimovs' article on their discovery of discontinuous permafrost has been rejected by both *Nature* and *Science*. A few months after I leave the park, the *National Geographic* article is published with the shocking headline "Some Arctic Ground No Longer Freezing—Even in Winter."[57] The Zimovs have managed to get discontinuity into print, albeit in a publication better known for striking imagery than for rigorous science.

The Zimovs represent a clash of knowledges across scales, worlds, boundaries, and time. They are gatekeepers of international research, yet their own research is often not considered rigorous enough to fully enter Western scientific circles. They provide media companies and journalists with a good story, with the implicit understanding they will be painted on a global stage as faintly ridiculous heroic adventurers—an identity they mobilize as a way to fund their rewilding experiment. In Chersky, they are both outsiders and insiders, at once subject to the smaller rhythms of Arctic town life and its accompanying unpredictable permafrost cycle and yet rich enough that they are able to leave the bitter winters behind.[58] They move across these permeable borders, sometimes with ease, sometimes coming up against tensions and frictions. Permafrost surrounds them, as it does all who come to Chersky, for whatever reason that might be. I, too, am surrounded, however briefly, by my awareness of permafrost at all times. As the ice on the Kolyma melts, and as the thaw depth probe sinks into the earth as if into wet cement, I am attuned to my own limited understanding of the warming going on around me and how it changes when I leave the islandness of Chersky.

The material ambiguity of this chapter has served to highlight how permafrost is at once both an earthly substance and an earthy one. So often, however, this earth is discarded in favor of Earth; permafrost thaw is funneled through Western globalizing processes, scientific singularity, and apocalyptic fearmongering.[59] The homogenization that takes place as a result of considering both the

Anthropos and its demise as singular and bounded *things* renders the permafrost as a singular object divorced from its patchiness and unpredictable freeze/thaw patterns. The luxury of imagining apocalypse as a future, probably far-off event necessarily accompanies the status of settler colonist, as does the hubris of imagining oneself as the titular (white) Anthropos. As Heather Davis and Zoe Todd state: "The Anthropocene . . . is really the arrival of the reverberations of that seismic shockwave into the nations who introduced colonial, capitalist processes across the globe in the last half-millennium in the first place."[60] In this sense, the apocalypse has already happened for millions of colonized people.[61] Introducing discontinuity into Anthropocene discourse recognizes that apocalypses can be small; apocalypses can be multiple; apocalypses can be historic or ongoing; apocalypses are political and contextualized by power. While the popular outreach from the likes of the Pleistocene Park and scientific media outlets acts as an attempt to connect people to a substance they have likely never come across, such a broad spotlight comes at the expense of more specific material entanglements. Emphasizing a singular earthly experience suggests a singular (and shallow) globe and a singular Anthropos with the ability to control its destiny and "save the world."

Such an emphasis must be resisted. Placing apocalypse in permafrost—specifically, Siberian permafrost—necessitates an engagement with the specificity of Soviet science, history, and geography and the ways in which the heterogeneity of the local and the *now* generates ruptures in the narrative of potential human extinction in the Anthropocene. When the anchored Soviet histories of the town of Chersky and its brutal beginnings are recognized, alongside the more localized engagements with permafrost and climate, the curated rendering of permafrost as a global carbon sink begins to disintegrate. When planetarity is imagined as earth rather than as a globe, and the jagged depth and shifting strata of permafrost are taken into account, the shallowness assigned to permafrost through mapping and modeling practices is afforded greater detail. Both of these interventions are discontinuous ones that recognize the dynamism and unpredictability of permafrost in the Russian Arctic. The careful and localized permafrost science done at the Pleistocene

Park and elsewhere is hugely important for its contribution to our understanding of the climate crisis, but it is similarly important to recognize that this type of science does not—and cannot—tell the whole story of the permafrost. Permafrost is much more than a singular scientific object (albeit a divergent one) of study. Permafrost can be ancient, but it can also be new. It can be a doorway to the underworld, or it can be an ecosystem for Sakhan horse grazing. It can give life to ancient viruses, or it can destroy the livelihoods of Sakhan fishermen. These permafrost apocalypses are ongoing and surprising, comprising pasts and futures; they demand attention and unsettle the perceived linearity and boundaries of time. Instead of defining an unruly earth by what it is *not,* imagine what it is and what it can be, what it *does* instead of what it has done *to* it. Paying attention to different responses emerging from the dynamism of permafrost freeze and thaw—the tusk hunter now spending his summers digging through mushy ground, the reindeer herder warning of evil spirits escaping, the displaced yak discovering a new sort of earth beneath its hooves—reveals not only a discontinuous earth but also a discontinuous Earth.

2 Ice

I arrive in Yakutsk at almost five in the morning after a hellish twenty-four hours stuck in Novosibirsk's airport. My flight was delayed over and over, the holdup at first attributed to "bad weather," but as the hours ticked by, people began to hear the news, and we found out that a plane had overshot the runway in Yakutsk and crashed. Thankfully there were no casualties, but the aircraft was destroyed and the airport closed until its stricken carcass could be removed. In the following days, I discovered that the cause of the crash was the airport's failure to de-ice the runway—a baffling decision, given that Yakutsk is the coldest city in the world and winter begins in September. It is now October, and it is snowing for the first time as I emerge bleary-eyed from the plane. The owner of the apartment I will be staying in has kindly turned out at this hour to pick me up, and we drive through the deserted streets as the snow muffles everything around us. I was last in Yakutsk five months earlier, when the heat and the dust and the wind had choked everything they touched; now I am about to experience the coldest winter of my life. Yakutsk is the largest city in the world built on permafrost; its buildings and infrastructure are designed around this fact, with thick metal girders sunken deep into the ground to hold up the tower blocks (simply balancing a full building on top of permafrost would cause slumping, as the heat from inside the building would melt the ground below). Yakutsk still has no bridge over the river Lena to connect it to a road because the erection of a structure that could last in such unpredictable conditions would require a massive feat of engineering.[1] For six months of the year, the only way in or out of the city is by plane; the rest of the time, winter blankets everything and people get around on "ice roads."

An ice road on the river Lena, Yakutsk. There is no bridge to Yakutsk across the Lena because it is so difficult to build one on permafrost, so the city is cut off from everything but air travel during the summer months; in the winter, the river is frozen enough to drive on.

I have come to Yakutsk to spend time at the Melnikov Permafrost Institute, a quiet place seemingly stuck in a Soviet time capsule, a world away from the very hands-on immediacy of permafrost science at NESS and the Pleistocene Park. Founded in 1960 by the pioneering engineer Pavel Melnikov, the MPI was at the forefront of permafrost science for much of the late Soviet era. Melnikov spearheaded the first attempt to bore groundwater holes into permafrost, a project that vastly improved structural engineers' understanding of how to build on frozen soils. During his thirty-year tenure as director of the MPI, Melnikov oversaw groundbreaking permafrost research in several areas: cartography, chemistry, geocryology, ecology, and engineering. The funding for this research was seemingly endless.[2] The current staff of the MPI are proud of their founder, and his office is kept as it was at the time of his death in 1994, as if he might one day return.

Today, the MPI is anything but pioneering. Dr. Alexander Fedorov, who has kindly agreed to mentor me during my stay and allows me

The office of the founder of the MPI, Pavel Melnikov—kept exactly as it was before his death in 1994.

to share his office, explains that the money simply dried up once the Soviet Union collapsed. His name appears on many academic publications, but most of these author credits have come from his acting as a gatekeeper or a guide for researchers from abroad who bring their own expensive equipment and big budgets. The scientists at the MPI are lucky if they can align their own fieldwork with that of these visitors and borrow instruments; if not, they are mostly stuck. Alexander has been experimenting with drone surveillance, flying over thermokarst polygons and comparing the pictures to photographs from thirty years ago. The drone is the most advanced piece of equipment the institute owns, but it cannot compare to the technologically advanced (and very expensive) remote sensing instruments available to scientists with huge fieldwork grants. There is a sort of frustrated resignation to Alexander's tone as he explains: "Usually we can just take pictures. In Russia we don't have access to these expensive instruments. It is impossible to take these measurements. It is impossible . . ." The traces of the former Soviet Union make themselves known in the CV gaps of previously prolific scientists.

Alexander is putting the finishing touches on a new permafrost map of Sakha, having also made the original map thirty years earlier.[3] The logistics involved in accurately mapping an area the size of Sakha (more than three million square kilometers and spanning three time zones) would require equipment and labor far beyond the reaches of the MPI. Much of Alexander's cartographic information is secondhand, gleaned from photographs, remote sensing models, scientific research undertaken by visiting scholars, even word of mouth; his physical mapping capabilities are restricted to a few sites around Yakutsk. He has lived in Sakha all his life, speaks Russian with a heavy Sakhan accent, and knows every fold and contour of his map intimately, even if he has never been to most of the mapped area in person. Down the corridor, his son is studying for a PhD in permafrost cryology. One day, Alexander rushes into the room and pushes a plastic shot glass full of bright-orange caviar into my hands, insisting that I try the local delicacy; he tells me he grew up being fed caviar by the spoonful. Another day he tells me we are ditching the office; it is the anniversary of the October Revolution, and we take to the bitterly cold streets to join others in a march under red banners to the ubiquitous Lenin statue. He lectures sometimes in Sakhan, trav-

Alexander Fedorov of the MPI surveys the permafrost map of Sakha that he completed in 1991. He finished an updated version of the map in 2018 while I was a visiting scholar at the MPI.

eling to some of the more remote villages in the area in an attempt to spread the word about climate change and permafrost thaw.

Alexander's story is just one of many to be found on the permafrost. The towns and villages clustered around Yakutsk are mainly home to Sakhans, many of them still practicing traditional animal husbandry and farming. Go farther north and you will find Indigenous Siberians who herd reindeer or horses. Some of them remain nomadic, while others live in villages along the Kolyma River: Eveni, Chukchi, Yukaghir.[4] Go to the west and you will find Nenets living on the Yamal Peninsula. The reindeer have their own stories, traversing vast tracts of tundra and navigating the various seasonal hardships: swarms of maddening mosquitoes in the summer, deep snow covering the tasty moss in the winter. Their blood produces a sort of ethanol during the coldest temperatures, sending the creatures into a drunken stupor, and Indigenous elders drink the reindeer blood for its nutritional properties. Reindeer can also be found in the Pleistocene Park; Nikita purchased a number of them from a group of herders up the Kolyma, transported them to the park on the barge, and released them. Down the road in Chersky, a former Soviet economic hub for shipping, most residents are ethnic Russians who remain despite the strong trend toward out-migration since perestroika. Chersky's cultural center houses artifacts and stages events for Indigenous Siberians. During the summer months, the husbands of the women who run the center travel deep into the tundra to hunt for mammoth tusks, which can be sold on the Chinese luxury goods market for a hefty profit.

These stories detail acts of responding to and living with a difficult substance; sometimes they intersect, and sometimes they do not. The people and animals involved are affected by permafrost thaw in different ways, and they understand permafrost thaw in different ways. Situated knowledges are built up over time by different groups, who are now contending with the fact that these knowledges are not really working anymore.[5] Permafrost is changing and disappearing, and the ways of life that have historically been curated by permafrost are similarly changing and disappearing. Can this be called an extinction of sorts? The stories I tell in these pages do the work of steering the definition of extinction away from normative claims. They are stories of friction, of liminality, of confusion; they disrupt assumptions of linearity and challenge scientific truths. Permafrost—and, indeed, the Arctic as a region—is known in multiple ways, and this means that the effects of permafrost thaw are experienced and responded to in different ways. It is not enough to acknowledge that a particular Arctic culture might go extinct; rather, permafrost extinction is a discontinuous process that produces new ways of knowing and living with tundra, in nonlinear and fractious ways. By telling just a few of these stories, by gathering together these patchwork pieces while acknowledging that they may never fit together or be complete, I hope to enable a deeper knowledge of permafrost.

Sakhan Refrigerators

I begin with a tunnel. Permafrost tunnels exist throughout Sakha, acting as refrigerators without the need for electricity. Known as *buluus* in the Sakhan language, they are a traditional Indigenous method of cold preservation and can be found dotted all over the tundra in varying states of use and upkeep. Some permafrost tunnels house more than food: the tunnel at the Pleistocene Park stores permafrost cores and other fieldwork samples; the tunnel at the MPI contains ancient seeds and bones as well as, for any tourists who drop by, gaudy ice figures of Father Frost, the Russian Santa Claus, and his daughter. There are personal tunnels for storing milk and fish, and there are vast, multiroom tunnels that act as archives and laboratories. It makes sense to take advantage of a cost-effective means of cold storage that exists in abundance below your feet; in fact, most people don't even think about it. Lena Sidorova, a Sakhan anthropologist at the North-Eastern Federal University, tells me that she grew up with a *buluus* and assumed this was the only way to keep food cold:

> My family, my father's family, we had this tunnel. Our grandfather's family have a tunnel. All people had such a tunnel. That's why it's absolutely normal. It's very interesting because it's something like using of permafrost, but I thought that *all* people in the world, they had such permafrost—because I was a child! In England, you have such tunnels too, you have to because you need a refrigerator! And when we moved to the city, I see [refrigerators] for the first time. . . . In the city they don't have this tunnel, and that's why we bought refrigerator, big white thing, and put it in the kitchen. So everything, very ordinary things, that's why when people ask us about the permafrost, it's not something unusual. All people had this tunnel, and we put in water from the lakes, ice, and meat, fish—everything that we need frozen.

The point at which people begin to notice their permafrost tunnels as something unusual is when things start to go wrong. Recently, these tunnels have started to melt, to subside, to flood.[6] The permafrost tunnel at the park is built on high enough ground that (the Zimovs assume) it will never flood, until it does just that. When the waters of the Kolyma subside enough, Nikita orders his workers to drain the tunnel using a motorized suction pipe. Once this task is completed, we take a boat to the tunnel to assess the damage and find that although the draining has been a success, the river water has left behind a thick film of ice coating the permafrost walls. Walking through the tunnel is like walking on an ice rink in the dark, and many of the tunnel's contents are now stuck fast to the walls and ceiling. Nikita brings along a spear to dislodge the bigger items, but some of the smaller objects will remain frozen in the ice indefinitely.

These snapshots of permafrost tunnels generate a simple question: What is different from before? Why are permafrost tunnels important *now,* as they are beginning to melt or subside or flood, or are being replaced with refrigerators? The MPI boasts that its permafrost tunnel is a fail-safe option, superior to other cryopreservational facilities because it does not require costly and possibly unreliable electricity.[7] Can this still be considered true? Lena's assertion that the absolute banality of permafrost became the subject of discussion only when the permafrost began doing things that were unexpected suggests this is a jarring progression. She says: "We never cared about permafrost. . . . If you are born in such a cold climate,

The drained permafrost tunnel at the Pleistocene Park after it flooded because of an unprecedented amount of meltwater in the Kolyma. Many of the items inside are stuck to the frozen walls and ceiling.

it's just weather, it's just life. It's the model of the world."[8] This is how people understand the permafrost in a practical way, as a way of keeping their food from spoiling and as a place to store safe drinking water. They also understand how to build houses on top of permafrost. These are tools of basic survival. The work at the MPI during the Soviet era focused mainly on the practical challenges of building on and manipulating permafrost to make it habitable for humans. The building of entire cities such as Yakutsk on top of permafrost represents a remarkable feat of engineering. It seems impossible that I might forget, in my status as a wide-eyed visitor, that permafrost is everywhere; it makes itself known to me constantly through warped and potholed roads, through the fat gas pipes that worm their way across the streets, and through the icy blasts that issue from underneath tower blocks.

Now, as permafrost thaws in strange and unpredictable ways, saying "It's just life" no longer works. The model of the world for the people of Sakha—the world of relatively stable and predictable permafrost, so ordinary that there is no word for it in Sakhan—no longer exists. Lilia Vinokurova, an Arctic historian at the Humanitarian Institute in Yakutsk, joins Alexander and me for tea and offers an explanation:

> So for a long time, permafrost investigations were more focused on technical and practical issues, not the social. You know, maybe it's connected to global climate changes and concerned with the forces of melting. . . . Ten years ago, one old lady in one small Yakutian [Sakhan] village told me: "When the eternal ice melts, where are we going to?" It was very interesting for me—so many of the residents thought that permafrost is permanent, and it isn't.

Permafrost is not permanent. "Life" is not stable. Life as a category has been troubled and critiqued thoroughly as a response to the compartmentalizing of structuralist thought in modernity, and the move to account for matter in terms of its dynamism and vibrancy renegotiates previously shored-up definitions and divisions that place life in direct opposition to nonlife.[9] Elizabeth Grosz states: "Life is not some mysterious alternative force, an other to matter, but the elaboration and expansion of matter."[10] The qualification of matter as an inert substance that the vital processes of life can control is a fallacy based on traditional scientific principles of a nature/culture dichotomy. The creation and renegotiation of life are inextricably bound up with the inorganic and the material forces that work between the two domains.

What does this mean in terms of extinction, which in its basic form accounts for the end of lives? The concept of extinction is very rarely applied to anything nonliving, and even when it is—most often in terms of languages or cultures—the examples are funneled through human social life. What would it mean for permafrost to go extinct, and what knock-on effects might be generated by the permafrost's *potential* to go extinct? That the permafrost is thawing, changing the landscape and becoming something else entirely, also changes the situated knowledges that have built up around living on and with permafrost: that it is (relatively) solid, that it is (largely) predictable, and that it is, indeed, (ostensibly) permanent.[11] As the preservational capacity of permafrost erodes as a result of global warming, the boundaries of what makes life, or what gets to count for life, are reconfigured. The permafrost tunnel, in its role as a cryopreservational facility, represents a convergence of nonlife supporting various forms of life—an example of how human attempts to regulate temperature are intermingled with the inhuman forces of freeze and thaw. When permafrost begins to thaw, and in doing so threatens the permafrost tunnel and what is kept inside, temperature becomes a cryopolitical issue in which definitions of survival and extinction are challenged by the dynamic potential of ice.

The Antipermafrost and Extinction in the Permafrost Tunnel

Ice and time are inextricably linked, both materially and socially—or, as Juan Francisco Salazar puts it: "Ice is in a permanent process of becoming time."[12] *Becoming* suggests movement, the exact opposite of inertia. In her book *Liquid Crystals,* Esther Leslie engages with the heterogeneous yet fractured nature of ice in all its forms. Her prose is full of the language of matter: slush, frost, flow, jelly, vortex, drift, creep, shear. Permafrost is but one of these, and it does not make an appearance until her conclusion, in the form of what Leslie calls "the mushy zone"; her point, ironically, is to render ice ephemeral, liminal, and certainly not permanent. She draws attention to the "implied immobility of the frozen ice, those huge blocks of solid water stuck in what seems to be another time and space, one without motion or sound—though polar ice reacting to changes of temperature may, in fact, creak, screech, howl and whine, and matter is never still."[13] The dynamism of less-than-frozen ice balances the sublime and the horror on a keen edge, a precipice that threatens an irredeemable decay and disappearance. Capturing ice through coring and freezing is an attempt to reverse the possibility of its retreat, any recession taking with it not only the knowledge frozen within but also time itself. To talk of a mushy zone is to recognize the temporal shifts generated by melting, by being frozen one moment and liquid the next, each material state producing different modes of life and survival.

Permafrost cores, like ice cores, are subject to this temporal mushy zone.[14] Permafrost cores might be classified as inorganic objects, but they contain layers and layers of organic matter that has built up over time—they are snapshots of permafrost as a mass. Unlike in temperate regions, in areas of permafrost organic matter becomes trapped by sediment and frozen before it has time to decay, thus any carbon that would normally escape and rejoin the carbon cycle is locked in. This carbon remains preserved in stasis until the permafrost thaws, when the microbial life awakens and begins belching out carbon dioxide. Permafrost cores offer glimpses of this layering process, of sediment upon sediment containing vast stocks of organic carbon as well as other substances, such as mercury (toxic to organic life). It is the particular linear framing of the cores that provides permafrost scientists with data regarding past climate events and the trajectories of climate conditions for supported life across time. The cores are used to predict carbon fluxes, temperature changes, and bacterial metabolic rates.[15] They are essential scientific objects, enabling researchers to analyze historical changes so that they can understand patterns of freeze and thaw as well as how permafrost behaves and shifts. Permafrost cores promise continuity, both of time and of knowledge.

When the permafrost tunnel at the park floods, piles of cores are stored in there—season after season of permafrost data. They are wrapped in plastic, but Nikita is not sure if they might have survived. "It was super hard for two weeks of extremely hard work for five people to get three hundred frozen cores, so it's painful," he tells me. "If we will not be able to save them it will be very sad."[16] The potential destruction of the cores has immediate ramifications for future permafrost research and for the scientists who have put their trust in the permafrost tunnel; stories abound from scientists working at the poles of cores and other samples that have ended up as puddles because of overzealous border guards or faulty freezing equipment.[17] In addition to demonstrating the integrated nature of doing science with unpredictable materials, such mishaps have wider and more abstract implications for permafrost life. What happens when permafrost cores and, by extension, the permafrost itself melt away into something that can no longer be described as permafrost? If nonlife provides the conditions for life to emerge *through* the material, what happens when the material changes, or when it disappears? Not only do these questions trouble the notion that permafrost is an inert, controllable substance that might preserve and produce life indefinitely, but they also suggest that its dynamic quality, through the forces of thaw, compels us to consider that life might retreat, or never even *become*.

Zooming out for a moment to consider the planetary reveals how the climate markers of the Pleistocene and trapped traces of the carbon cycle are contained within a permafrost core. Carbon, the building block of all life, represents the biggest threat to human survival on the planet, externalized as the enemy to be brought under control.[18] Locked in ice, carbon is safe; it does what nonlife is *supposed* to do and provides the correct conditions for life to flourish. When permafrost thaws, carbon as enemy is released and begins to threaten life. It is this imbalance, almost a tug-of-war between life and nonlife, that produces the mushy zone. The permafrost tunnel can no longer be trusted to preserve and produce life; instead, it destroys histories and strata. Its retreat and mutability generate an excess of carbon, which shakes free of its expected role of maintaining life and instead unsettles, warps, and creates other forms of life that might

Permafrost cores in the Pleistocene Park's permafrost tunnel. Permafrost cores offer scientists an excellent way to track changes across layers of deep time.

be unwelcome to humans. This has implications for the ontological positioning of permafrost as nonlife and the ways its encroaching *impermanence* destroys the notion of the boundary frozen in time. The potential for permafrost to become extinct produces further anxieties around surviving with—or, indeed, *without*—permafrost. Ice cores capture the fundamental paradox of ice and time in the Anthropocene: they are temporal objects, tethered to a material and linear continuity of deep time that collides with the quickness of the "now" that melting lays bare. The knowledge gleaned from ice and permafrost cores of carbon spikes and climate-altering events proves that such things could happen again; that permafrost is currently thawing—and taking the cores along with it—proves that such a thing is happening *now.*

Thinking through permafrost extinction in this way, as a threat to life, is just one of many possible approaches to understanding what is happening to permafrost and the Arctic in the Anthropocene. I have more stories to recount of the ways permafrost is generating discontinuity through thaw. Many of these stories are not about permafrost at all; rather, they are revelations of the effects of permafrost thaw and retreat. They are stories where the banality of permafrost comes to matter in surprising ways, not necessarily because the permafrost ceases to *be* banal (indeed, much permafrost thaw goes unnoticed), but instead because its new rhythms of slow thaw and shift renegotiate the continuity of ontologies, lives, materialities, temporalities, and spatialities that have relied on permafrost staying put (or, at least, having predictably cyclical dynamic qualities). The effects of this renegotiation are often jarring, violent, and surprising—the punctured tundra skin releasing a cacophony of demonic bacteria, the slumping of thermokarst revealing the corpses of mammoths, the flooding of permafrost tunnels. These occurrences unsettle the ways permafrost has (historically) been known, while the reverberations of thaw might challenge how extinction is defined at a time when living with or understanding permafrost is being discontinuously remade. The permafrost in the following stories is confounding, boundary transgressing, insidiously creeping and retreating. Permafrost's unpredictability renders it impossible to stay with or to follow closely, and as a result any stories that emerge from this space are similarly difficult to keep up with; what I recount, therefore, is already long out of date, long out of time. The point is to anchor a brief moment of permafrost life before it ebbs away and becomes mush, but also to take into account what might emerge from these absences.

The Permafrost Law and Inhuman Subjectivity

A permafrost tunnel is but a snapshot of a wider process happening all over Sakha. In October 2017, the Sakhan government announced it would be implementing a "permafrost law" designed to protect and safeguard permafrost in the future—or, more accurately, to safeguard the people and buildings whose survival was dependent on permafrost behaving as it should.[19] Historically, permafrost was considered a difficult substance to build on, but once it was conquered, it could generally be relied on to stay put. Permafrost in the twenty-first century is a different matter entirely, or, as the director of the MPI put it in 2018: "Nothing that was built on permafrost was built with the expectation that it would melt. We must prepare for the worst."[20] The permafrost law is a preventive measure aimed at protecting the permafrost to ensure the safety and survival of the people who live in the permafrost environment; it includes restrictions on building, industry, and agriculture that are intended to limit the damage to permafrost already being wrought by climate change. The principles of the law are set out as follows:

> The priority of protecting the life and health of the population in the interests of present and future generations; ensuring favorable environmental conditions for human life, work, and rest; prevention of irreversible consequences of the degradation of permafrost as a result of geocryological processes; . . . transparency, completeness, and reliability of information on the state of permafrost and its changes, forecasting the sensitivity and stability of permafrost landscapes.[21]

Yakutsk is the largest city in the world built entirely on permafrost. Its infrastructure (including roads, buildings, and pipelines) is suffering the effects of increased thaw. This street displays many of the telltale signs, such as buckling and veering.

Alexander is on the advisory board for the new permafrost protection law, which is still in the consultation phase with the Sakhan government. He confesses to me that the law might be too little too late. One day I come into the office to find him watching news footage of a collapsed building that had crushed a gas pipe, causing a fire. All over town, many of the buildings sport deep cracks, with girders lurching off to the side and pipes splayed at awkward angles. My own apartment steps veer terrifyingly. Most of the repairs that have been done are mere stopgaps; the city lacks both the funds and the novel expertise to respond to such a rapidly deteriorating situation. Lena takes a more cynical view:

> People are talking more about the melting, and at first, it was about roads. Our administration, they just simply didn't manage to repair the roads, and I don't know, maybe some of them just stole money? But the reason was "Oh it's just the permafrost! We can't do anything, it's the permafrost. It's climate. It's the nine months of winter!" Something like that. It was some using of permafrost for administration, for Russian administration, for making excuses.

While the practical and far-reaching effects of such a law remain to be seen, the implementation of a law that protects humans *through* the protection of inhuman matter generates some interesting questions around what gets to count legally, how such a law will be enforced, and the slippery modes of subjectivity and agency that become inscribed legally into the geopolitical (and cryopolitical) system. In particular, the ways subjecthood gets renegotiated through an acknowledgment of the dynamic qualities of permafrost, and how the law comes to understand the inhuman through the creation of a permafrost subject, further redefine what extinction is in the Anthropocene.

Kathryn Yusoff calls for a greater interrogation into what she calls "geologic life," which considers the "mineralogical dimension of human composition" but also how attending to the inhuman helps to trouble the homogenized Anthropocene subject—the white capitalist man—and unearths the subtending forces of a more open geological subjectivity.[22] What conclusions might be drawn from the permafrost law if it is considered through a lens of geological subjectivity, particularly in regard to the future? While the permafrost law makes no attempt to categorize permafrost as imbued with life, it nonetheless ascribes to the permafrost object a sort of irritating liveliness that lawmakers call on their constituents to take seriously as Yakutsk and other landscapes of Sakha begin showing signs of instability. There is a certain antagonism in the way that government officials blame permafrost liveliness that is beyond their control for the state of Sakhan roads while also acknowledging the need for a law to protect the permafrost. What the permafrost law attempts to do is to legally inscribe a permafrost subject that is created through the actions of others elsewhere. The law does not attempt to lay out mitigation strategies for climate change, acknowledging that the continued heating of the Arctic is an inevitability; instead, it sets out strategies and protections from the effects of permafrost thaw. Or, as Alexander's coauthor on the consultation draft, Eduard Romanov, put it in a 2017 interview: "It is extremely important to draw the attention of society, deputies and politicians to the existing problem of permafrost concerning each of us. After all, our 'tomorrow' is already predetermined."[23]

A permafrost subject is produced through both the global manifestations of climate change and the increasingly apparent localized effects of thaw: the collapsing of permafrost tunnels, the buckling of roads and buildings, the evisceration of important agricultural lands. The creation of a permafrost subject is a way to mark out the renegotiation of life through the specificities and differences of inhuman matter—in this case, thawing permafrost. While the law does not recognize multiple permafrost ontologies, it does point to multiple geological materialities that both unsettle the notion of a homogeneous planet and human species and draw attention to the differential material forces that work with (and against) different forms of matter. It is the degradation of permafrost that threatens permafrost life, through the excess carbon that escapes and through thaw's destruction of the stability of the ground. In the same interview quoted above, Romanov stated: "Permafrost is insidious. It has a local character and a surprise factor."[24] That this is the understanding of the permafrost law suggests a move away from the

Soviet-era permafrost science of manipulation and control and toward a recognition of the dynamic agency of nonlife that is inextricable from the survival and the promotion of permafrost life. But it is also a question of protection and harm. The permafrost has achieved recognition in law *through* its instability and vulnerability, and the law is very much contextualized by permafrost's ability to do harm and be harmed by climate change. While this attention is to be welcomed, the unspoken caveat of offering subjectivity and protection *now* suggests that permafrost's "natural" state—of supposed inertness and permanence—renders it objectified once more. Elizabeth Povinelli calls this distinction out through her discussion of a river, stating: "We need to overcome the division of the lively and inert, the vibrant and listless. . . . Maybe she wants to gradually decline into the inert. Maybe she is tired of all this becoming."[25] If the permafrost stopped making itself known through buckled roads and burst pipes, would the law renege on its protection?

How the law will be enacted and enforced remains to be seen. It might be that legal protections could end up cleaving to a sort of "doubling down" on categorizing permafrost as controllable, so that the law merely requires adjustments of current engineering practices such as the insertion of girders deeper into the earth or increased checks on structural stability (much of the consultation on the law so far points toward prevention and detection). But the shift from using and manipulating permafrost landscapes for human habitation toward a more sustainable "living with" underwrites the preservational properties of the inhuman for the promotion of life, with the text of the law making the connection between the health and well-being of the population and the stability of permafrost landscapes.[26] Recognizing the instability of the permafrost also highlights how this new permafrost subjectivity is bound to the whims of freeze and thaw. It is important to note, however, that the permafrost subject is not a creation of this law but an entity that is in constant flux through the shifting materialities of permafrost, an entity that is now beginning to be recognized in governmental structures. The retreat of permafrost—and, indeed, the potential loss of what permafrost *is* to Sakha—becomes curated through the law as an extinction: the recognition that what permafrost livelihoods need to survive is the protection of the permafrost landscape. Yet viewing permafrost through the lens of human lawmaking practices cannot account for a subjectivity that is *not* bound by the promotion of human life; the permafrost is becoming extinct, perhaps, as Povinelli might put it, on her own terms. Redefining extinction so that it incorporates the nonliving must also account for—or at least acknowledge—the impossibility of knowing what form such an extinction might take and the unpredictable effects it may have.

Does Permafrost Breathe?

The effects of permafrost thaw can bubble up from below in surprising ways. In August 2016, dozens of people living on the remote Yamal Peninsula in Arctic Siberia were hospitalized as a result of anthrax infection. Experts attributed this strange occurrence to a single dead reindeer whose bacteria-riddled carcass had been frozen and preserved in a permafrost tomb for more than seventy-five years. After a particularly hot summer, some of this permafrost had thawed, warming the carcass and releasing what has come to be known as "the Siberian plague." A living herd of reindeer, already weakened by poor grazing due to the unseasonably warm temperatures, disturbed the spores through their foraging and became infected. More than two thousand reindeer developed the bacterial infection and passed the pathogen to their human handlers; many of the reindeer perished, and specialist troops trained in biological warfare were deployed to the region to manage the situation. A twelve-year-old boy died. Hundreds of surviving reindeer were slaughtered to prevent further infection, their carcasses burnt on pyres and the ground washed with bleach.

This incident was a mere blip in the global media: poor provincial Arctic folk menaced by something not unlike the zombie parasite in the Arctic-set thriller TV show *Fortitude*. The story made for a shocking read over morning coffee and then was quickly forgotten. But for the Indigenous Nenets who make their lives in this now-ravaged part of Siberia, the anthrax outbreak caused irreparable damage to their relationship with the tundra.[27] To the Western world, bleaching the ground and burning infected reindeer carcasses seemed to be logical steps to prevent such a thing from happening again. Russian officials have since stated that in the future thousands more of the surviving reindeer may need to be euthanized as a precautionary measure—the herds are too densely compacted, they say, as if their teeming liveliness is almost too successful.[28] The Yamal Nenets are the last Siberian Indigenous group to maintain an almost fully nomadic lifestyle, and living and moving with reindeer is fundamental to their autonomy; the elders drink the nutrient-rich blood of their cull, children pick out sinew with their teeth, and reindeer hides are used for clothing and toys.[29] Carcasses are buried, but the usually solid permafrost ground means these graves are shallow and subject to exposure through unexpected thaw. At a press conference, the head of Russia's public health watchdog agency, Anna Popova, stated: "Children were infected as a result of traditional customs."[30]

It is difficult to ignore the unpleasant undertones of victim blaming in this context. The Nenets had practiced their traditional customs for centuries without incident before this infection, which was caused by an unseasonably hot summer that thawed the permafrost and exposed the dormant anthrax spores. Such so-called unseasonable summer temperatures, however, are becoming the norm. In Chersky, I talk with Nataliya, an Evenk woman who now lives in town and helps run the center promoting Indigenous cultures. She lives in town because her daughter has a congenital heart condition and needs to be close to medical care, but some of her family still keep reindeer. Unlike the herds of the Nenets, the Evenki reindeer herds have dwindled as a result of Soviet assimilation and suppression of Evenki culture, low pay for herders, and disinterest from younger Evenks. Climate change is yet another hurdle, as Nataliya explains:

> There are few remaining—we have only two herds now. There were ten before, then eight, then less and less. Young people are leaving and don't want to work. . . . It's harder, there was a lot of snow this year. Quite a few [reindeer] died. My sister is a reindeer herder. It's very hard. We would find calves dead from sinking in snow. . . . My sister and her husband, they went to the forest and immediately got out because there is more snow

Chersky is home to people from several Indigenous groups. Nataliya, an Evenk woman, shows me an Evenki reindeer fur headdress at the town's cultural center.

> there than on the tundra.[31] There is little help and the salary is very low.

Traditional knowledge is being lost. Nataliya stresses that this is why the cultural center exists, to keep these practices alive somewhat, at least in memory. She talks of the "Russification" of the town's children; although she has taught the Evenki language to her own children, she notes with sadness that most speak only Russian. There are times, however, when Indigenous ethnic identity might matter less. Although anthropologists have painstakingly sorted and categorized Siberian ethnic groups, Lena reveals a rather different dynamic that makes these groups and identities much more fluid:

> So they actually do not need this identification, because they can call themselves "tundra people." It's really for their lives—it's enough. It's absolutely enough. . . . They call themselves tundra people, not like a congress or community, because for a long time, they live in one area, they have one way of life, and they live together and they work together, and because it's a cold climate,

> they have to be together always. They do not care who is who. When it's cold and you're hungry, it doesn't matter!

So Evenki, Eveni, and Chukchi are all tundra people. Nenets are tundra people as well, even though the Yamal Peninsula is a long way from Chersky. What binds tundra people is permafrost, along with the generations of knowledge they have built up from learning to survive in such an unforgiving landscape. A situation such as the Nenets faced in 2016 demands an inquiry into how this tundra identity is underpinned by climate change and thawing permafrost. This requires an ontological repositioning—a way of knowing permafrost beyond its scientific categorization. When Popova blamed the anthrax epidemic on the Nenets' adherence to traditional customs, she was speaking from a position of Western medicine, of identifying the points at which the pathogen infected human bodies and suggesting cures or preventive techniques. Even equating the incident with climate change–induced permafrost thaw cleaves to a certain Western scientific doctrine, in that permafrost is treated as a passive substance within which danger lurks. Tundra people, conversely, are all too aware of the dynamic agency of the ground beneath their feet—not through any sort of dichotomy between life and nonlife, but through their attunement to the ways in which the permafrost moves through the seasons, often sensitive, surprising, and occasionally vulnerable. The Yakuts speak of the three spirit realms: the upper world, where the gods live; the middle world, where humans live; and the underworld, which houses evil spirits and demons. While there are fire spirits and earth spirits and sky spirits, there is no permafrost spirit. Tundra, instead, acts as a carpet—a ceiling—between the middle world and the underworld, demanding respect. As Lena explains:

> The word is used with a deep sense: tundra. So it's "because it's tundra." It's tundra. Tundra is everything. Tundra means . . . tundra is some zone or place like another. Something another. I heard several times: "Tundra is very sensitive." I didn't perceive this sensitivity very well before I went to the tundra this summer. So if you live in tundra, you have to be very mobile. You cannot

> stay in one place because just imagine, if you stay on the earth, in the ice—it's ice! If you stay, you are warm. And if people stay in one place, they are warm, and of course they have this carpet preserving the earth, the soil, so they cannot stay in one place.

This relationship to tundra has parallels with the fuzzy boundaries and entwined notions of place, personhood, and kinship found in Inuit relationships with glaciers. As Julie Cruikshank describes in her book *Do Glaciers Listen?*, for the Inuit, glaciers are "both animate (endowed with life) and animating (giving life to) landscapes they inhabit."[32] Permafrost occupies a similar space of almost symbiosis; tundra people know how even tiny imbalances of the earth can result in danger or even death. This knowledge is not rooted in permafrost monitoring or focused seasonal tracking but comes from an understanding of the permafrost as a sort of vascular communication network that may respond to poor treatment with hostility.[33] The Sakhan scholar Nikolai Sleptsov-Sylyk details this understanding in his book *The Breath of the Permafrost*, in which he chides his fellow Sakhans—who, while not native to the tundra, have made their homes on the permafrost for centuries—for turning their backs on traditional ways of living with permafrost and moving to cities. He emphasizes that any damage done to the permafrost causes wounding that, if not allowed to heal or blessed by a shaman, will allow for evil spirits to escape the underworld and cause all manner of problems:

> Our ancestors lived in the bosom of nature and were hypersensitive people to manifestations of the subtle world. . . . Of course, nature is capable of self-healing, but the process is very slow by the standards of human life. In permafrost, nature is very vulnerable. We quickly destroy the surface of the earth, and nature has no time to tighten its wounds, and in turn, these injuries also affect us.[34]

Sleptsov-Sylyk may be unfair in generalizing about Sakhans, but his book provides an insight into a permafrost ontology that is sorely lacking in other academic texts. The land physically breathes, emitting foul odors from wounds that have not been allowed to heal; unpleasant atmospheres linger around certain "zones." To build a house on top of permafrost, one must find a spot where the breath is warm, avoiding areas of colder air and ideally disturbing the ground as little as possible. Thawing causes a hollowing of the land and attracts evil spirits; the warmth of a household built on permafrost can allow demons and other spirits to gather and wreak havoc. The permafrost, like skin, can heal itself, but it must be given enough time; the speed of current rates of permafrost thaw means that wounds are left open to fester, and different and conflicting timescales clash.[35] Although Sleptsov-Sylyk does not equate faster permafrost thaw with climate change, he makes it clear that the turn toward the greater convenience and comfort afforded by city life, and the relinquishing of old knowledge, has resulted in the permafrost taking its revenge on those who would disrespect it.

The normative definition of extinction in the Anthropocene renders Sleptsov-Sylyk's work as a cultural idiosyncrasy, but to deny the breath of the permafrost—and the very real signs that indicate this breath is failing—is also to deny the centuries of knowledge and tradition of the people who dwell within it. Knowing tundra is an embodied practice, and one that recognizes the agency and dynamism of a materiality that the singularity of science would categorize as lifeless.[36] The wounding of the permafrost is a lack of something, a gaping hole through which evil spirits can escape, a space of retreat and a rendering to mush. Demons and diseases surface because of the disruptive material state of too much thaw, the newly porous boundaries of permafrost skin or surface tension laying bare the fallacy that only life can experience extinction. As with many calls for decolonization, recognizing that other ontologies exist beyond Western science opens up spaces and perspectives that understand permafrost thaw as not only the loss of tundra but also the end of particular lifeworlds that have flourished *with* tundra.[37] The "breath of the permafrost" reveals the gravity of what happened on the Yamal Peninsula: pouring bleach over the tundra is akin to pouring bleach into a flesh wound. Perhaps the anthrax spores have succumbed for now, but the permafrost will continue to thaw, year after year, and as it does, new relations and practices will emerge in response.

Permafrost as Commodity and the Hunt for White Gold

One of these new practices has produced a booming black market in Sakha. Every summer, groups of men, mostly Indigenous, head deep into the tundra and do not return until the autumn. They are after the ivory of mammoth tusks, once locked deep within frozen ground but now becoming more and more accessible as the permafrost thaws.[38] There are officially sanctioned tusk hunts, but most of the hunters operate illegally, using high-powered water cannons to blast through mushy permafrost and make tunnels. These tunnels are dangerous, susceptible to subsidence and mudslides, but the deeper the hunters go, the greater the chances they will strike white gold. A single tusk can sell for eye-watering amounts, traveling from tusk hunter to buyer to Chinese markets, carved into ornaments, shaped into jewelry, displayed in homes as markers of status.[39] Hunters who find big, intact tusks can expect payouts far beyond the average monthly salary of Sakha (around five hundred U.S. dollars); hunters who find nothing (as most do) usually lose money. Tusk hunting is a gamble, but the allure of striking it rich is enough to drive many hunters to return summer after summer.

These men have been raised on stories and mythologies that involve mammoths. Many such mythologies have emerged through observations of the material landscape; for example, the Sakhan belief that mammoths were like giant rats that tunneled through the permafrost arose from that fact that thermokarst features are reminiscent of subterranean burrowing.[40] Similarly, the Evenki understanding that mammoths—*khele* in the Evenki language—were aquatic stemmed from the presence of rivers and other bodies of water that cut through permafrost banks; so efficient were the *khele* at carving up the landscape, they eventually sunk beneath the ground.[41] Although the belief systems of Russia's Indigenous peoples have been damaged by generations of cultural destruction and assimilation wrought by the Soviets, there remains a general understanding that disturbing a mammoth carcass is a bad idea.[42] In thinking similar to Sleptsov-Sylyk's regarding how Sakhans ought to make amends to the permafrost they violate, it is widely believed that actively searching for a mammoth body bodes very badly for the hunter, particularly if he breaks through the permafrost to find it. If a mammoth is found organically—an increasingly rare occurrence—the burial site must be consecrated with offerings.[43] Hunters who now spend their summers searching for mammoth bodies to dig up face a choice between angering the spirits and making some badly needed income. When I ask Nataliya how she and her husband navigate this tricky decision, she responds:

> Nowadays nothing is considered a sin, but before, even if you found [mammoths], you need to make—I remember my mother would weave a necklace out of round beads and would drop it in the same place [the burial site], to stop bad spirits from coming out, otherwise someone dies. Nowadays no one's afraid of anything. My husband left and I wasn't on time to make the beads for him, so if he finds a tusk . . .

I ask if he will take the tusk regardless. She answers: "Yes, he wants to. Of course, I forbade him. But nowadays they don't see things that way." Nataliya is afraid for her husband. He has disobeyed her and gone to the tusk hunting site anyway, without beads for protection. The hunting grounds are often extremely remote, as the hunters must explore areas farther and farther away from settlements because of the decreasing availability of mammoth bodies. Each year, more tusk hunters join the groups or attempt to find new sites. Far from law enforcement and the disapproval of their families, and frequently bored, they often turn to alcohol and violent behavior, sometimes resulting in death.[44]

However, tusk hunters are much more than lawbreakers who scavenge in isolated corners of the tundra—they are also indispensable sources of information for others seeking mammoth carcasses: scientists hoping for new opportunities to understand mammoths, to sequence their genome and study their phylogeny, and, of course, scientists working on de-extinction, hunting for a perfectly preserved strand of DNA from which they might clone a mammoth. The Mammoth Museum in Yakutsk hires tusk hunters who deliver any well-preserved prehistoric creatures to the museum laboratories. One afternoon, I am having tea with the head of the laboratory, Dr. Semyon

Grigoriev, when two hunters burst into the room to report a find. Their photos are blurry, taken in the dark with a mobile phone, but they show a part of a baby mammoth in a decent state of preservation.

As Semyon and the hunters chatter in Sakhan about transporting the carcass to the museum's giant freezer, I am struck by the irony of thawing permafrost producing the conditions that encourage tusk hunting. Here, the cryopolitical and ontological boundaries of what constitutes life become further blurred, paths and motivations crossing and clashing.[45] The dead mammoth becomes an intersection of worlds: there is agreement on the creature's status as dead, but a Sakhan tusk hunter might feel the gnaw of anxiety as he rips a tusk from the fleshy face of a mammoth he has disturbed, whereas an ecologist might feel a spark of grief (or perhaps hope!) as she surveys the corpse of a long-extinct species. The thawing or blasting of permafrost produces new economies of luxury goods crossing borders, while simultaneously paving the way for genetic research that might very well result in *Jurassic Park*–style mammoth petting zoos. This is a cryopolitics of disturbance, made possible by the digging and water-hosing of thawing permafrost—its discontinuity arising here not from climatic patterns but through the flows of capital that pay for the bodies and hacked-apart pieces of mammoths shipped all over the world. Permafrost is the treasure chest that hides the potential riches of tusks and genes, and is hollowed out with little regard, its accelerated thawing revealing new spaces of speculative accumulation, renegotiated as lifeless, inert dirt to be washed away, any belief in its lively and vengeful potential ignored in favor of making money. These are often agonizing decisions. Many of the tusk hunters have no choice but to join this exploitative new industry, as their traditional labors and pastimes are thwarted by terrible pay and climate-induced disruption. The homogenizing force of the global capitalist project disregards any objection to or anxiety about its juggernaut of extraction and instead attempts to bulldoze the multiplicity and surprise of permafrost into something much more malleable. As much as there is hope to be found in the possibility of "capitalist ruins," a cursory glance toward the Arctic suggests that a new wave of extractive prospecting will reconfigure permafrost as a last-gasp cache of capitalist commodification.[46]

Tusk hunting is indicative of a form of permafrost life that has emerged from permafrost's retreat, displacing other permafrost lifeworlds; as traditional Indigenous economies such as reindeer herding and other forms of agriculture are threatened by Arctic heating and increasingly poor economic returns, tusk hunting has moved in to fill the gap. The reasons and motivations for this are complex, involving more than solely a response to thawing permafrost; they comprise difficult histories, legacies of violent suppression, and complicated and slippery identities. Woven tightly through the histories of Siberian Indigenous groups is colonialism's systematic devaluation of their traditional knowledges and ways of life, first through Cossack brutality and then through Soviet cultural homogenization.[47] As much as the women at the cultural center identify as tundra people and mourn the loss of their old way of life, they are also proud Russians, as evidenced by their celebration of Russia Day, when Chersky residents come together to affirm their love of the Motherland and unveil a giant tapestry of Vladimir Putin's face. The point is that embodied tundra living inscribes a fractious and shifting process of identity that is inextricably bound to the dynamism of permafrost. Lena is rightly dismissive of the way many settler colonists romanticize the lives of Indigenous peoples, pointing out that the reason tundra people are particularly attuned to their environment is because they have to be to survive: "You are not 'close to nature,' you *live in* nature! It's not something close to nature. Because when people say, 'Ooh, Indigenous peoples are close to nature'—it's like some sort of poem!"

That some Indigenous people disregard their beliefs about mammoth gods to hunt for tusks reflects a process as discontinuous as permafrost thaw itself. While many Indigenous groups in Siberia are facing—or, indeed, have already experienced—the extinction of their lifeworlds as a legacy of the twin disasters of colonial violence and climate change, there are many instances of subversion and resistance. Lena tells me that many Indigenous people embody several identities—an Evenki might pose as a Chukchi, for instance, to represent the interests of all tundra people on government panels. Tusk hunting indicates a similarly fractious identity that wrests agency back from barely fit-for-purpose welfare provision and a Sakhan government unable to halt the climate crisis. While tusk hunting is indeed motivated by flows of global capital stimulated by a lust for status objects, it can also be seen as a sort of subversion of, or at least reaction to, the difficult landscape of thawing permafrost. The point is that when it comes to the idea of a permafrost extinction, there is no road map for what the response or aftereffects might be—we can only know that they are discontinuous.

Unearthing Multiplicity through Stories

I have told several permafrost stories that document how thaw and material retreat interject and make possible the emergence or suppression of new and unexpected forms of life. There are many more stories to be told, of course, and I have left many out, but it soon becomes apparent from this gathering of brief snapshots that it is impossible to align them in a way that will help us to understand permafrost as a singular and whole knowledge object.[48] The permafrost tunnel, with its preservational properties, safeguarding the genetic building blocks of life on ice, is increasingly subject to melting and flooding. The high-rise buildings in Yakutsk, built within the framework of a controllable and inert permafrost, are soon to be under the jurisdiction of a permafrost protection law. Indigenous cosmologies view the permafrost as a skin, and to wound it is an act of violence that brings death and misfortune. Disturbing a mammoth carcass, the remains of an underworld god, risks danger, but it might also mean riches for the tusk hunter who takes a chance, using illegal water cannons to blast aside the permafrost, like shooting bullets into flesh. None of these ontologies can live comfortably beside the others, but the increasingly pressing effects of global climate change are drawing them all closer together—certainly not in a way that suggests a neat concatenation, but rather in a raw, jagged, and discontinuous way. These nodes of tension seep across scales and temporalities, infecting and destroying the notion that the Anthropocene concept encompasses all humans. The future apocalyptic event becomes impossible at the sight of reindeer burning on pyres, the bleaching of the tundra, mammoth carcasses rotting in the sun. Similarly, the neatly layered strata so key to a geological definition of the Anthropocene cannot be pinned down when the earth is in such flux, undulating with the materialities of freeze and thaw, or blasted away through extractive ecologies of disturbance.[49] If the Anthropocene indicates the dominance of the human as found in the fossil record, these traces of human activity and existence are similarly erased by the thawing of ice.

How, then, to understand permafrost extinction in the Anthropocene, at the juncture of such disparity? Permafrost, in one form or another, covers almost a quarter of the Northern Hemisphere's land surface; inasmuch as it is discontinuous, local, and specific, it is also important to acknowledge it as a vast region of the planet. I argue that we must place permafrost when it is so often contextualized as a "nonplace"; remoteness, particularly Arctic remoteness, occupies a sort of timeless quality in the minds of people living far away.[50] Ksenia Tatarchenko states that narratives of remoteness create "a violent topology of the distant and disconnected"; the very remoteness and thus homogeneity of permafrost representation renders it disconnected from the rest of the world but fails to capture the heterogeneity found within its depths.[51] Permafrost sits at the difficult intersection between the global and the intensely local, spanning an entire hemisphere but behaving discontinuously throughout. This traversal of scales and subjectivities comes from the forces of freeze and thaw that imbue the permafrost with its unpredictability and dynamism. To conceive of a permafrost extinction means challenging its global categorization—it is unlikely that all permafrost will disappear—and instead allowing for smaller, more banal extinctions that register on a local level, or might even pass unnoticed, while also recognizing the global effects of permafrost thaw on the planet's climate.

Permafrost, while certainly never the inert and controllable substance those Yakutsk builders believed it to be (or did they?) in the early 1990s when they were constructing cheap high-rises in a "Wild West" economy, has begun "waking up" through accelerated thaw as a response to anthropogenic climate change. This has implications for the planet through the surging of greenhouse gases that will occur if excessive thawing continues; while a dramatic permafrost catastrophe is unlikely, the effects of increased levels of carbon in the atmosphere will be felt globally. The global implications of permafrost thaw can also be found in the flooded permafrost tunnel, the emergent anthrax bacterium, the likely futile attempts to safeguard a populace from harm through legal measures. They are contained within this disconnected topography of permafrost, which draws together a disparate "tundra people," but the sum of its parts does not make the whole; there can be no defined permafrost object when its material boundaries are so fuzzy. The heterogeneity of the permafrost object is located in Anthropocene ruptures—nodes of whiplash demonstrating that the homogeneous configuration of the Anthropos is a fallacy, the logic of planetary redemption is flawed, and the notion of a singular future apocalyptic event is a myth.

There are instances in which the materiality of permafrost, or, more accurately, its material retreat through thaw, bleeds across ontological boundaries through a process of infection, erasure, and redirection. While a retreat of physicality itself, the spaces of absence that are formed once permafrost becomes something else produce new engagements and responses to the *lack,* be they anxiety, fear, opportunity, frustration. An ontological repositioning of this magnitude involves placing one's feet firmly in the mush, allowing the meltwater to trickle into one's boots; this is not an ethical positioning but rather an invitation to be uncomfortable. Thinking similarly of fluidity and depth, Philip Steinberg and Kimberley Peters proffer a turn toward a wet ontology, drawing on the swell and liquidity of oceanic spaces as a way to dismantle geography's traditional commitment to fixity and bounded place.[52] While I have taken great pains to emphasize the dynamism of permafrost, in that its permanence is a misnomer and its materiality unpredictable through freeze and thaw, I also want to point to the ways in which it retreats, changes, or vanishes altogether *through* this dynamism. A wet or oceanic ontology might account for the ways in which boundaries are transgressed and volumes breached, but it tends to cleave to a certain material *overflow,* in which there is often an excess of watery liveliness. There is an allure and a thrill to the oceanic that does not exist in talk of permafrost, either because of ignorance about its existence or because of a general disinterest in permafrost's muddy, subsurface banality. As Lena said: "We never cared about permafrost." Essentially: we never cared about permafrost until it was no longer there; we never cared about permafrost until its absence became noticeable.[53]

This is how we might conceive of a permafrost extinction, as a response to the banal ceasing to *be* banal, although never too much. If extinction is understood as a rupture—a shuddering end to something that produces knock-on effects—then it might be tempting to think of a permafrost extinction as a shocking, violent event. But that may not necessarily be the case. While permafrost is still dynamic through its slowness and its impermanence (and occasional violent slumps), at the same time its retreat and absence represent a refusal to enter into some encompassing vitality of life that so often centers the human. We should notice the permafrost precisely because it is largely unnoticeable—notice the ways in which its receding disrupts the carefully crafted fallacies of linear time and life, continuous earthly materials, and singular ontologies. It is impossible to know permafrost fully *because* it is disappearing, becoming erased, its surprising emergences and revelations generating new ways of looking at the world, however briefly. The point is that permafrost as nonlife is not separable from life and its dependence on support and coevolution *with* permafrost, so that any permafrost extinction will necessarily have aftershocks for permafrost life. These may also manifest as an extinction—the inability of Sakhans to build their houses on a bit of land with no bad spirits, or the violence of pouring bleach onto the skin of the Earth—while the retreat or end of a form of permafrost life might open up spaces for new forms to emerge. What is important here is not only the different ways of knowing and living with permafrost recounted in these stories but also the ways in which certain forms of permafrost knowledge and life are privileged over others through regimes of politics and power. Those spaces that emerge from permafrost retreat are filled almost exclusively by those curated by capitalism, by extractive designs, and by assumptions of ontological singularity. The discontinuity produced by permafrost extinctions must similarly contend with the buttressing of hierarchy—not everyone experiences permafrost thaw the same way. But when it comes to living with permafrost, in all its different forms, the spaces its thawing leaves behind are anything but empty; rather, in an Anthropocene underpinned by a desperation to maintain a continuous and singular lifeline of extraction and mastery, the multiplicities found in those absences are disruptions that reveal a discontinuous Earth.

The Lives Underfoot

There is a meteorological station some distance down the Kolyma—actually, so far down that it juts out into the Arctic Ocean, on the final headland before the autonomous region of Chukotka. The only inhabitants are two couples who live and work together for periods of two years at a time, followed by a six-month break. During the two-year shifts, they monitor the weather and climate patterns, transmit the data to a central facility, and speak to very few people aside from each other. A ship comes once a year to deliver supplies, but occasionally Nikita makes the four-hour boat trip to bring the families fresh vegetables and other perishables like cheese and eggs. I ask to go with him on one of these trips, and he agrees, but with a warning: "Fifty percent of the time I think I'm dead, it's on this journey."

I understand this only when we leave the sheltered safety of the Kolyma tributary and head out onto open water. We are thrown around bodily as our little boat bounces across the waves, occasionally lurching and coming near to flipping. I'm worried for the eggs. We drink a shot of vodka to calm our nerves. When we arrive more or less in one piece, the snow is glistening in the June sunshine; it is still too cold for it to melt. As Nikita stacks the supplies onto a snowmobile to drive up to the station, I take a walk along the shore. Piles of fire-blackened cans and other discarded items are strewn about—there is no waste collection here, of course, and the solidity of (this) permafrost prevents rubbish from being buried. Eventually I reach a wooden hut with a large metal cross placed on top. Twisted coils of barbed wire snake through the grasses, the remnants of a perimeter fence, and the rotting remains of buildings. Ambarchik was not always a meteorological station; Ambarchik was once a gulag.[54]

Thinking about permafrost lives will always be an incomplete process; the discontinuity of permafrost in the Anthropocene produces the necessary unpredictability to trouble notions of life and nonlife, and the ways in which these slippery categories are renegotiated by

Ambarchik is accessible only by boat, and the only people who live there are two couples who monitor the weather station. This wild, desolate headland was a gulag in the 1930s, and remnants of that history still remain.

thaw. However, it is important to remain aware of the physical presence of bodies underfoot. The knowledge that the bodies of hundreds, if not thousands, of people are beneath my feet at that moment (Russia's refusal to acknowledge the extent of the gulag atrocities means that accurate numbers are impossible to know) forces my attention to the very brutal nature of the earth's makeup. Geology, as Kathryn Yusoff asserts, is never neutral; the Anthropocene is what it is because of the millions of Black bodies forced into mines and slavery.[55] So much of Russia's wealth (and potential wealth) is found in permafrost, its gains in permafrost science and engineering resulting mostly from a lust for precious minerals rather than for knowledge. The bodies—the former *lives*—underfoot inscribe this permafrost landscape; they inscribe its very materiality through the diamond mines of Chukotka, the gas fields of the Yamal Peninsula, the opening up of the Northeast Passage for freight.

There are those in Russia and beyond who welcome a melting

Arctic.[56] These cryopolitical negotiations bank on a world that is getting hotter, an Arctic where temperatures are rising more rapidly than anywhere else on the planet, and a thawing permafrost that allows for the extraction of fossil fuels that have been, up until now, unattainable. This is a cryopolitics that intends to preserve and embed extractive and violent practices, that seeks to maintain divisions between life and a lifeless inhuman matter. None of the stories I have recounted are any better than others, or come any *closer* to a full understanding of permafrost, but some of them are indeed presented as though they are better by those who have a particular stake in permafrost behaving a certain way. What is preserved is the homogenized human form—white Western man—with difference categorized as lifeless and inhuman through further violent practices.[57] But the unpredictability of permafrost thaw means it is impossible to preserve and prolong life in a certain stasis. What life is privileged in the permafrost tunnels—the human lives, the "good" bacteria, the endangered DNA and seeds—is unsettled by the refusal of permafrost to stay where it is, to do what it should. Instead, there might be a retreat, a vanishing through the forces of thaw, or the emergence of "bad" forms of life through anthrax spores or evil spirits. Or life may never emerge at all, the DNA of mammoths decaying gently in the Arctic sun. Permafrost renegotiates what survival and extinction mean in the Anthropocene, disrupting cryopolitical strata and challenging what gets to count as life. Ironically, the paradox of the Anthropocene emerges once again: it is anthropogenic climate change that provides the conditions for permafrost to become its most surprising and threatening. The seemingly banal renewal of human life atop the permafrost goes unnoticed until it is interrupted by permafrost's refusal to do what is expected.

My aim in this chapter has not been to offer some universal definition of permafrost extinction—indeed, permafrost is often not even called permafrost at all—but rather to point to the blurring of boundaries between life and nonlife, and to what emerges or retreats within such fuzzy spaces. It is through universalizing that permafrost becomes lifeless, a splodge on a map or a set of statistics on a graph; its inscription through a narrative of isolation and remoteness renders it a realm of the placeless. I am loath to call permafrost "tundra" because not all permafrost *is* tundra and not all tundra is permafrost. Permafrost retreats, decays, becomes something else through freeze and thaw, producing discontinuous effects that are rooted in place and identity. These discontinuities are extinctions insofar as they produce other things and become something else; they are endings, and the spaces they leave behind are new beginnings. They are similarly experienced in different ways by different actors—many more than I have briefly documented in this chapter. Permafrost is revealed to me in different registers and through different configurations wrought by language, by sensing, by walking, by sight and smell and touch, often partially obscured by breakdowns in unshared knowledge, unshared language, unshared culture and perspective, hidden by its isolation and subterranean makeup. Does permafrost breathe, then, as glaciers listen? Does permafrost hear, speak, know, act, feel? When I touch the walls of the flooded permafrost cave, they are entombed in cold, shiny ice; when I hold a ball of thawed-out soil between my fingers, it is bouncy and malleable, like putty. How does this substance alter *me,* as I traverse faint paths across the Arctic tundra or trudge through knee-deep snow in Yakutsk? I feel my body lurch to the side as I climb the steps to my apartment block; I feel the sponginess, the slipperiness, of thawing permafrost beneath my feet at Duvanny Yar. Cruikshank rightly cautions Western researchers against aiming to "capture" Indigenous knowledge through data collection and quantifiable measures.[58] My own relationship with permafrost is enriched by the people I speak with and by my brief immersion into the tundra landscape, but I can only ever tell these stories through limited words and insufficient language. Displacing the dominant Western ontology and epistemology requires listening but never appropriating, being open to difference but never enforcing. Like Lena, who realized that she could not perceive the tundra's sensitivity until she spent time there, I understand that my own perception is made legible by my experience of slowly becoming attuned to permafrost, but only ever partially, only ever discontinuously.

3 Bone

Summer solstice. The longest day of the year, when the sun reaches its zenith across the Northern Hemisphere—particularly noticeable in the Arctic, where the skies remain as bright at 3:00 a.m. as they are at 3:00 p.m. Each year, Sakhan towns throw a solstice festival, called Ysyakh, dedicated to the "greeting of the sun." In Chersky, this event is a chance for the town to come together in celebration, with singing, dancing, and special foods, along with traditional sports such as Sakhan wrestling and reindeer antler lassoing, as well as a game invented by Nikita that involves tossing a heavy disk as far as possible. Every year local men try to beat Nikita's record, and every year he retains his title. There is a friendly (and sometimes not so friendly) rivalry between the locals and the Zimov dynasty, who are considered outsiders by some of the townspeople. Not only is their work with the Pleistocene Park looked on with bemusement, but also their relative wealth, displayed through the likes of brand-new Land Rovers and custom-built houses, occasionally draws envy from the Chersky residents who struggle to make ends meet. "People eat our animals!" Nastya exclaims angrily during one of our drives into town for supplies. She recounts an instance when one of the locals drunkenly boasted about shooting and eating one of the musk oxen from the park. The truth of the man's story has not yet been confirmed, but the Zimovs are following up with the local police—another act that has hardly endeared the family to the town. But Nastya is resolute: "They don't understand what we are doing here—they think animals are for eating."

It is possible to be sympathetic to both parties in this case. To be part of a dwindling population in a dying Arctic town—where food is scarce and prices are sky-high, where job prospects are mainly seasonal and poorly paid, where Moscow seems a very long way away—can only be a difficult existence. An ox lumbers in front of you, and you have your gun. Wild animals are rare in these parts, the tundra mostly empty and barren save for the hordes of mosquitoes that feast on your blood. It takes but a moment and you have a supply of meat to last you weeks. You know the musk ox must belong to the Zimovs' park, but it is still a wild animal, and animals, after all, are for eating. For the Zimovs, though, every animal loss is felt keenly. Their musk oxen have traveled many miles by rickety boat, plucked from a lonely herd on isolated Wrangel Island, last stronghold of the mammoth. Other animals have come from even farther away—northern Mongolia, Lake Baikal, and even Europe, not to mention the herd of baby bison sitting in a pen in Alaska awaiting the logistical and bureaucratic go-ahead to make the trip to Siberia. All of this costs vast amounts of money—money that the Zimovs, while personally rather comfortable financially, must raise through the successful running of the station, alongside crowdfunding campaigns and bids for awards from initiatives like the TED Audacious Project. All this to return the ecosystem to its Pleistocene glory.

To understand just what this means we must return to the summer solstice—a cloudless and blindingly sunny day in the Siberian Arctic. Nikita announces at breakfast that we will visit Duvanny Yar: "Everyone who visits must take a trip to Duvanny Yar," he jokes. "It's my main tourist attraction." We will also call in on Leonid, the climate change–denying fisherman who likes to serenade his guests with Russian guitar songs. We pack two bottles of vodka—one to drink during the ride and one to share with Leonid—and set off in the boat on the four-hour journey. We eventually reach Duvanny Yar, a little worse for wear, and are confronted with a giant permafrost bank that juts out as an overhang, the waves of the river biting away at its underside. Chunks of permafrost are crashing into the water, rivulets running down the muddy bank face in the heat of the sun. Nikita points out ice wedges and ancient tree roots as we shunt along to the shore and tie up the boat. Duvanny Yar is a product not of climate change but of the slow and sustained erosion of the bank by river currents that have exposed the raw face of permafrost, which then thaws in the summer months. It is here that

Sergey developed his ideas regarding the spatial extent of the mammoth steppe ecosystem and the density of animal life that existed there during the Pleistocene era. He did this through the discovery of thousands of bones in the mud of Duvanny Yar; having once been buried and locked inside permafrost, the topsoil had steadily crumbled over time, unearthing and revealing a veritable graveyard of ancient skeletons.[1] Nikita speaks of being unable to walk down the beach without tripping over a hip or a skull. Bones are a little more difficult to come by now, but he is reasonably confident we will find something if we dip our hands into the thick mud.

We do. After about five minutes of searching, we come up with a bison hip ball joint and part of the shoulder blade of an ancient horse. It is difficult to grasp how old the mud-caked piece of bone is. Around forty thousand years old, Nikita estimates. A piece of prehistory the likes of which I have only ever encountered before in museums, mounted behind glass cases, certainly never held in my hands. As the sun beats down on the permafrost below, I think about how many of these bones might be beneath our feet. How deep and dark must this burial site be? From the mathematical logic of this vast deathscape emerged the specific parameters for the Pleistocene Park; the density of skeletons found at Duvanny Yar meant that the Zimovs could estimate that thousands upon thousands of animals once roamed the tundra—the complete opposite of the ecological desolation found today. Bison, yak, horse, woolly rhinoceros, mammoth—most of them now, if not officially extinct, at least long gone from this region. As we wander back to the boat, Nikita scrambles up to a sandy bank and begins to kick at it wildly, sending dust flying up in choking clouds. "I am making faster permafrost degradation! I am collapsing the soil! I am burying the ice wedges!" he shouts through the dust. "You're like the mammoth," I say. He slides back down the bank. "Better," he replies. "A human mammoth."

There is a long lineage of scientific work done at the Duvanny Yar site, and many important discoveries regarding permafrost behavior and the Pleistocene remains found there have emerged from fieldwork conducted on this patch of slippery mud: the carbon dating of ice wedges, Pleistocene silt analysis, regeneration of thirty-thousand-year-old fruit tissue, and even the discovery of allegedly

Nikita, the human mammoth, mimics the destructive power of large herbivores in trampling permafrost soils.

forty-thousand-year-old worms, which were taken back to a laboratory in Moscow and gently warmed until they began to wriggle around.[2] The Zimovs' own work at Duvanny Yar involved a complex series of mathematical equations and the help of a Pleistocene bone expert who could identify all the skeletal remains plucked from the permafrost in a designated area.[3] According to Nikita:

> We actually managed to get actual data for numbers of species in this ecosystem, and we got some quite impressive numbers. So, for every square kilometer of this mammoth steppe ecosystem, even in the coldest regions, there was living one mammoth, five bison, eight horse, and fifteen reindeer. . . . If you look around you, you will see two to three thousand animals around. So, it's a huge number. Right now, if you will climb up the hill in the place where I am, you can climb up every day you will not see anything. Maybe lots of mosquitoes—that's it. Oh, what a rotten world!

For the Zimovs, knowing the exact numbers of animals found in the mammoth steppe ecosystem is integral to their Pleistocene Park experiment. This is a long-extinct ecosystem, with many of its main players, including the titular mammoth, also extinct, but the main purpose of the park has never been conservation or the restoration of animals; rather, the Zimovs' objective is to re-create prehistoric ecological processes in a landscape that, to them, seems lacking somehow. As proponents of the "Pleistocene overkill" theory, which places the blame for the decimation of the mammoth steppe on human activity rather than on climate change, the Zimovs view their role as that of custodians of a project that aims to put right the past wrongs of early humans. This is a common refrain among those who claim we are now in the sixth great extinction—that is, the current background rate of species extinction far exceeds what is deemed "normal"—and the first one to be caused (both directly and indirectly) by human activity. While not officially considered a geological marker for the Anthropocene, the sixth great extinction is, indeed, a result of it. In its current form, the Pleistocene Park might be seen as a confluence of nonhuman extinctions, converging at the point of different temporalities: that of a deep, Pleistocene past that saw the destruction of an ecosystem, which became a repository of bones, and the current rate of species extinction, a reminder that survival is a precarious thing in the Anthropocene. The bones found in the soils of Duvanny Yar and farther afield act as material anchors that hold past, present, and future loosely together in discontinuity.

How such temporalities come into play emerges from the way these bones are arranged and excavated to form a prehistoric imaginary of Duvanny Yar's geological and ecological past. The painstaking work of reconstructing the density of animal skeletons joins an assemblage of scientific work that bleeds out into the wider world, into the international scientific community, where it may take on other forms and produce other knowledges along horizontal and wider scalar planes. These knowledges create legacies of scientific understanding, building gradually on what came before, just like soil accumulating on top of bones. That permafrost is now thawing and that Pleistocene bones are now accessible to the average humanities

At the Royal Geographical Society, Nikita explains the density of animals at Duvanny Yar during the Pleistocene epoch. This calculation formed the basis of the plan for the Pleistocene Park.

scholar who might chance to dip her hand into the mud lays bare the fact that accumulation—of both knowledge and time—can just as easily be eroded. The very verticality of permafrost and its cache of preserved history demands attention; while the vast majority of engagement with permafrost takes place at the surface or just beneath (on account of its accessibility and dynamism), permafrost as a vertical space produces a different sort of legacy that forces different timescales into confrontation when that verticality is threatened by thaw.[4] The notion of legacy is tricky in a space like the Pleistocene Park; any rewilded animals necessarily collide with an ecological history buttressed by extinction and ruin. How the extinctions of the past inflect the (potential) extinctions of the future underpins the park's entire strategy of "saving" permafrost, but it also points to broader questions of subterranean relations, ecological ruin, and the discontinuity of time.

Permafrost Underground and the Verticality of Time

The majority of permafrost's dynamic activity occurs in the upper reaches of the soil layer. The seasonal thawing and freezing of the active layer is a necessary part of the "natural" permafrost cycle, allowing vegetation to take root and support an ecosystem. Lately, however, increased thawing and deepening of the active layer due to a warming climate have resulted in material, ecological, and hydrological changes to the permafrost body as a whole. The warmer the surface temperature or the thicker the insulating layer of snow, the deeper the summer thaw, and sometimes, as the Zimovs discovered in the summer of 2018, that active layer may not freeze again come winter. But the overall depth and material heterogeneity of permafrost matter a great deal here, a fact that is rarely noted in popular science and mass media reporting on permafrost. Some permafrost can reach depths of almost two kilometers, at which point it is so old and so far from the surface, it has remained frozen for millennia. Official data are scant, but in 2008, scientists dated an ice wedge in northern Canada to 740,000 years old.[5] Given that assumptions had been that all permafrost had disappeared during the interglacial warm period about 120,000 years ago, this was a momentous find, and one met with a degree of skepticism by other permafrost scientists.[6] But if the deepest, coldest permafrost had weathered a climate that was warmer and wetter than the current one, perhaps permafrost was more stable than first thought?

The subterranean planet is usually imagined to be mostly stable and unchanging, save for the odd earthquake or volcanic eruption.[7] The underground is often left out of geopolitical boundary demarcations, considered lifeless and quite boring to anyone outside of geology, inaccessible and homogeneous, something to be taken for granted as we conduct the much more exciting business of living on the surface.[8] There is a hazy, and perhaps uneasy, awareness of the subterranean, and it bubbles out occasionally through literature and film, through stories of terrifying beasts emerging from the depths, of the hubris of drilling too far down.[9] These tales warn against disturbing the relative calm of the underworld, which is quite content to leave us alone as long as we don't go prodding and prying and disturbing. These are not just spooky stories, either; most ancient and modern religions and belief systems contain references to a subterranean realm, either as the locus of an afterlife or as the domain of demonic entities. The Sakhan epic *Olonkho* tells of the journey of the hero Nurgun Botur the Swift into the underworld to defeat the evil spirits that are causing havoc on the surface; not only is this a rousing adventure, but it also encompasses the traditional belief system of the Sakhans, that there are three separate worlds—underworld, surface, and heavens—and any imbalance among them is the ultimate taboo. The notion of an imbalance is connected ontologically to the ways Sakhans live on the surface and engage with the tundra beneath their feet. Yet despite these specific cultural and cosmological impressions, the underground remains largely untheorized, or even unthought. According to Matthew Kearnes and Lauren Rickards: "In contrast to the assumed instability and impermanence of the worlds aboveground, the subterranean is read as . . . offering the promise of permanent stability across an almost unimaginable time horizon."[10] Similarly, Nigel Clark draws attention to this tendency and queries why "modern western political discourses have come to focus overwhelmingly on the organization of the *surface* of the earth rather than on its layering or multi-dimensionality."[11] The subterranean—in its material sense at least—is assumed to be both an apolitical space and a stable one.

Dig deeper into the realms of history and geology, and it becomes clear that this is not the case. The discovery and study of fossils and skeletons in the eighteenth century revealed a vast underground deathscape of deep pasts, a time before human existence, prompting a new understanding of the planet and its chaotic revolutions and extinctions. It was the arrival of this subterranean carnage at the surface that generated anxieties concerning the weight of history on the present and future: If these mass extinctions had happened to other species, might they happen again? In particular, paleontologist Georges Cuvier's analyses of fossil and carcass discoveries across Siberia drew attention to how certain histories and certain futures become constructed. Having derailed the budding discipline of Western geology with petulant smears and misinformation, as well as subscribing to deeply racist and often downright unscientific beliefs, Cuvier presented ideas around extinction and the remains of extinct species that were revolutionary at the time.[12] His catastrophist thinking about earthly "revolutions" causing massive and sudden upheaval to planetary life was not particularly original, but his work became very influential as a result of his dogged pursuit of his theory of a prehistory without humans (and his propensity to explain away anything that did not fit the theory). Cuvier's work had a significant impact on geological thinking, cementing the ideas that catastrophic revolutions are wholly natural and separate from human culture, and that the fossils and bones of extinct animals are "nature's own history."[13] In the face of this vast subterranean bonescape, life suddenly became much more dynamic, and much more precarious.[14]

Bones are powerful symbols that resonate widely across sociocultural practices, particularly those related to funerals, burials, and cremations; many religious ceremonies incorporate imagery of bones "returning" to the underground—ashes to ashes, dust to dust. Bones are also political indicators—of war, brutality, and colonialism, as well as the struggle for rights through such acts as the repatriation of bones, which serve as physical locators for mourning. The physicality of bones is important, too, as highly differentiated from yet inextricably linked to the flesh, which dissolves away after death, leaving the bones intact.[15] The bones of both humans and animals are trace signifiers of life and death, yet with an important spatial caveat: human remains are usually interred underground in graves, or at least hidden from view, whereas animal bones lie where they fall, producing a jarring uncanniness that inflects different temporalities and scales of death. That bones endure, aided by permafrost, for millennia, only to be found on the shores of Duvanny Yar by Russian scientists, is a ghostly memorial of both individual animal death and species extinction. "Bodies have the advantage of concreteness that nonetheless transcends time, making past immediately present," states Katherine Verdery, and the emergence of skeletons and bodies from the subterranean to the surface upends the notion that the underground—and the bones contained within—stays where it is.[16]

How, then, do these bones become actors themselves in such a complex space, and how do they produce future imaginaries of planetary redemption? Through an exploration of the interment, discovery, and excavation of the bones at Duvanny Yar, a picture begins to emerge of how the Pleistocene Park came into being, beginning life in the 1990s as a casual experiment by Sergey and then taking on greater weight as more of the bonescape was revealed and more of the permafrost retreated and thawed. This is a process of ruination, in which the material processes of decay produce both the ruins themselves and a by-product of affective spectrality that lingers at the site.[17] The bones at Duvanny Yar are ghostly registers that anchor the remains of the Pleistocene ecosystem to a landscape that no longer exists; they are agents themselves in the park's trajectory across its short existence, transgressing timescales as remnants of a deep past, but also of a shallow history. It is in this sense that ruins cannot be seen to be the remaining vestiges of things that are irrevocably gone or dead. The ecological ruins of Duvanny Yar and beyond vibrate and hum with past ghosts but also point to the very present and ongoing processes that generate cryopolitical fantasies of control over life.[18] The bones and the bodies of extinct prehistoric creatures are not merely things that once were, they are material actors that disrupt neatly layered strata and linear temporalities, producing discontinuities of thought and matter. Ann Stoler describes "to ruin" as "a virulent verb," stating: "We are as taken with ruins as sites that condense alternative senses of history, and with ruination as an ongoing corrosive process that weighs on the future."[19] It might be pertinent to dwell on the double meaning of "remains"—both the physical reminders of ecosystem collapse and animal death and the remaining thrum of haunting anxiety that bleeds across time, both shallow and deep.[20] The temporalities housed in permafrost bleed out and inflect how the Zimovs curate their project, the ruins of a prehistoric ecosystem acting as a locus through which they redefine *and* construct the past.[21] The physical remainders of past extinctions act as revolutions that threaten the supposed stability of the subterranean, drawing together timescales both shallow and deep.

Deep Pasts: The Super-terrestrial with(out) Humans

There is a tension here, then, in the Anthropocene subterranean imaginaries of stability versus revolution. As the Anthropocene Working

Group inches slowly toward a resolution, the issue and the mapping of subterranean strata take on greater prominence.[22] The underground becomes important, not merely as an extractive resource but also as a way to categorize and date both the destructive practices of the humans above and the deposits they leave in the ground, jagged and contested strata emerging through meetings, academic papers, and geological excavations. What exactly leaves its mark most firmly in rock, what produces the shiniest golden spike? Still, these debates concern themselves with relatively shallow strata, with the likeliest starting date for the Anthropocene being the 1950s, the decade of concentrated nuclear weapons testing. It is unsurprising that the cultivation of a human-centric geological epoch is concerned mainly with the shallowest of timescales, given the blip in time that human life on the planet represents. Valerie Olson and Lisa Messeri identify this disconnect between shallow and deep strata as an inner/outer dichotomy, in which the Anthropos is naturally concerned with its environmental milieu at the expense of the outer: "Other processes and things get bracketed out. . . . The Anthropocene concept is overdetermined by anthropic relations with inner environment and undetermined by anthropic relations with outer environment."[23] Their example focuses mainly on the extraterrestrial, but we might extend that focus to what could be termed the "super-terrestrial": earth and strata so deep that they are beyond human intervention or discovery. These strata represent outer epochs, stretching back four and a half billion years; the Pleistocene, an epoch that began two and a half million years ago, is a relatively young timescale encompassing the most recent glaciations, yet it still surpasses the epoch of human life by many millennia. To put it into perspective, the oldest human remains ever found have been dated to around 300,000 years ago.[24] These "outer" timescales of the super-terrestrial are very old indeed, almost unfathomable; they contrast sharply with the much younger timescale of human history. We spend our short lives in this shallow pool, in the writings and recordings of past scholars, in the self-centered bubble of this inner world. What, then, happens when these deep pasts, considered so separated from modern human agency and practices, reveal themselves and collide with the shallow histories of anthropogenic dominance?

Whereas golden spikes and epochal transitions are based on the vertical stacking of sedimentary strata upon strata, fossils acting as material temporal anchors for the linear progression of time, the thawing of the permafrost at Duvanny Yar represents the erosion of this time, the gradual decay and retreat of strata that reveals a deeper past. That the discovery of the bones at Duvanny Yar led Sergey to conclude that it was Pleistocene overkill, and not the more commonly accepted factor of climate change, that produced the shift from steppe ecosystem to tundra is key to understanding why the Zimovs believe they can restore the mammoth steppe.[25] The Pleistocene overkill theory, which asserts that the mass die-off of many prehistoric mammals, including the woolly mammoth, was caused by anthropogenic hunting and habitat destruction, remains controversial in many scientific circles, but Sergey and Nikita are adamant.[26] Sergey has published on Pleistocene overkill in prestigious science journals; he has conducted modeling and collected evidence to support his claim. Nikita, however, is a little more irreverent:

> There are many stories that prove this. We refer in our articles to the guy "Przewalski"—you know the Przewalski horse? So he was Polish explorer who was exploring Central Asia, Tibet, China, and Far East. . . . He was a big scientist, big name, and the most advanced person at that time. And in his diaries he was saying every day he was shooting like thirty kilograms of bullets. When he was driving on the barge he was just like shooting everything. So he was coming to this place where there was no people, to Tibetan plateau, and he saw thousands and thousands and thousands of animals and he killed half of them. Immediately. He was writing about how many animals he killed for no reason. And here, two or three centuries ago, there was a herd of reindeer—a huge herd of wild reindeer caught in the Kolyma. And people encountered that and they killed all of them—I think a hundred and fifty thousand? There was a huge mountain of them.

It is clear that Nikita and his father hold a dim view of the actions of humans in the past, and that they extend this opinion to early humans living alongside mammoths. Nikita goes on to explain that he and Sergey are not claiming that humans slaughtered every last creature, just that they killed enough to cause the ecosystem to change from a highly nutritious grassland to a scrubby tundra with very little to eat. The animals could no longer survive there. The discrepancy in mammoth populations is often pointed to as supporting evidence for Pleistocene overkill: that mammoths died out on the mainland a full four thousand years before they did on isolated islands like Wrangel appears to bolster the claim that humans were at least partly responsible for this extinction event. Or, as Nikita proffers casually: "If you kill a mammoth, you're the sexiest guy in the land, all the chicks are yours."

Nikita's flippant remark is indicative of something deeper that informs both his and his father's understanding of ecology and predation. It is an understanding that is found throughout much of rewilding discourse: that not only do humans not belong in nature, but they actively cause it harm.[27] Ironically, rewilding in the vein of the Pleistocene Park occupies a space in the so-called good Anthropocene, in which humans maintain control of the planet by removing themselves from a nature that will fix it for them. Whether the Pleistocene overkill hypothesis is correct or not is beyond my purview as a humanities scholar; I lack the required expertise to make a claim on either side of the debate (although it seems most likely that a combination of overkill and climate change is responsible for the extinction). However, my intention is not to interrogate this hypothesis scientifically, but merely to attempt to understand the logic and rationale of these two scientists who subscribe so fully to it. The discovery of the bones at Duvanny Yar meant that the Zimovs were confronted with a defunct ecosystem rendered visible by permafrost thaw; in their eyes, it is a product of humanity's destructive prowess.

This defunct ecosystem—with its vast expanse of prehistoric animal bones belonging to creatures that once shaped and maintained a grassy steppe—contrasts sharply with the scrubby tundra found today, which supports very few animal species. These bones do not invoke human memory—they are simply too old—but a sense of anxiety emerges when one is confronted with such massive evidence of extinction, the possible catastrophic futures wrought through permafrost retreat, and the release of greenhouse gases colliding with the tangible and material reminders of a lost past.[28] These animal remains gnaw at the physical reality that the working ecosystem of the mammoth steppe is no longer there and generate imaginaries around the potential consequences of ecological ruin.[29] But what also emerges from this encounter is a sort of imagined nostalgia—a pieced-together memorial collage of a balanced ecosystem lost to the destructive potential of humanity.[30] Much of the discourse around rewilding hinges on the desire of humans to right the wrongs of the past, and while the Zimovs profess not to care too much about the animals in their care beyond what they can do for the ecosystem, the landscape of extinction at Duvanny Yar presses upon what they do with the Pleistocene Park, and how the park gets constructed as a utopic space of salvation.[31] Whether it was Pleistocene overkill or climate change that ruined the ecosystem, what emerges from the deep past of the permafrost burial site is a fantasy that the Pleistocene Park might save the world not only from the anxieties of the present but also from a potentially catastrophic future. This sort of techno-utopic dreaming might join the geoengineering projects favored by proponents of a "good Anthropocene," but it also has its roots in twentieth-century Russian history.

Shallow Histories: The Soviet Union and the Genesis of the Pleistocene Park

To stand in the mud and hold a forty-thousand-year-old bone in your hands fosters a connection with deep time, but the shallower histories of Duvanny Yar and their effect on the Pleistocene Park's genesis cannot be overstated. Russia's current approach to ecology—and, largely, Sergey and Nikita's too—cannot be separated from the country's social and historical practices, both political and scientific. In particular, Russian scientists in the latter stages of the Soviet Union were influenced by the work of Vladimir Vernadsky, a geochemist whose oeuvre spanned the waning days of the Russian empire, through the Bolshevik Revolution, the Stalinist purges, and World War II. Vernadsky is currently well respected in Russia (and, consequently, less so in the Western world) for his ideas around the biosphere and for originating, along with another scientist, the concept of the noösphere, or the sphere of thought.[32] For Vernadsky, the noösphere represented the progression from an entangled ecological coevolution toward the recognition of humans as benign geological agents affecting Earth processes through scientific and technological progress, in the manner of intelligent stewardship. Consequently, some scholars have pointed to Vernadsky as recognizing the Anthropocene more than half a century before that term entered modern parlance.[33] In a 1945 essay, he described the noösphere as "a new geological phenomenon on our planet. In it for the first time, man becomes a large-scale geologic force."[34] What would have been quite jarring to Vernadsky, had he lived into the twenty-first century, is the notion that such anthropogenic power could possibly be a bad thing. These scientific ideals of benign human stewardship aligned well with Soviet materialist philosophies of utopia, and while Vernadsky showed no great love for Bolshevism, he was nonetheless spared the fate of forced gulag labor that befell many of his colleagues. His school of thought helped change the Russian approach to earth sciences from one that was largely descriptive to a more dynamic and experimental interrogation of the subtending properties of earth strata as generative of what he termed "living matter."[35] His ideas around the coevolution of life and nonlife have been key to modern Russian environmental thinking, and have also encouraged more practical directions for scientific research. In the 1930s, Vernadsky helped to spearhead the Soviet study of permafrost.

Fast-forward fifty years, and we might meet Sergey Zimov, a young geophysicist at the Far Eastern State University in Vladivostok. He would have undoubtedly engaged with Vernadsky's ideas, which were beginning to be reintroduced to curricula, and his more philosophical texts make reference to "Vernadskiy's law" regarding the coevolution of the biosphere.[36] The Pleistocene Park is based on these ideas, of ecosystem renewal and restoration involving cycles and turnovers, of the rewilded animals stimulating geological and ecological processes and vice versa. In his "Wild Field Manifesto," a sprawling mission statement advocating for the return of pasture, Sergey pulls no punches, condemning the destructive practices of humans that have removed them from their benign and symbiotic role in the ecosystem. It is here he deviates from Vernadsky's valorization of the technologically augmented human mind; while he is still a believer in "Vernadskiy's law" of ecological entanglement, his situatedness in a past and present that saw the collapse of the Soviet Union, the "Wild West" years of the 1990s, and the cumulative effects of anthropogenic climate change and ecosystem collapse has led him to a much more muted vision of humanity's maintenance of the noösphere. Sergey's utopia involves a hands-off form of stewardship, one of humans learning how to live as part of a system again, relinquishing absolute mastery and using their superior minds to educate themselves and reconnect with "wild nature." Or, as he concludes in his manifesto: "We don't have a gene of zealous masters of Earth; this, [as] with many other things, we have to learn."[37]

Nikita's milieu was rooted even more deeply in political and ecological turmoil. Sergey's pioneering work on climate change, and his sheer single-minded obsessiveness in setting up the Pleistocene Park, meant that Nikita spent most of his childhood during the 1990s without his father around. The decade following perestroika was difficult for many Russians, and for the Zimovs it was no different. "For my family, I can remember we were quite poor," Nikita tells me. "I remember, maybe it was '92, something like that—my dad was somewhere away and my mum saw the check for the monthly salary and she looked at the sum and she started crying." Nikita was largely raised by his mother, Galya; a respected scientist in her own right, she spent her PhD years in Kamchatka studying volcanoes. She is an astute and thoughtful woman, obviously highly attuned to her husband's often erratic moods and ideas, and she tells me of the early years living in Chersky, of bitterly cold winters and paraffin instead of electricity. I ask Nikita how the station and the park survived such economic tumult, given that every other Russian Arctic research station had ceased operating. He tells me his dad anticipated the fall of the Soviet Union years before it actually happened and squirreled away fuel and materials, repurposing old buildings and motors. "The whole town, and the station as well, were professional scavengers," he explains. "We were eating off the dead body of the Soviet Union."

These shallower and perhaps more sociopolitical histories become strata themselves, both entangled in deeper earth pasts and surpassed by them. As Nikita neared the conclusion of his mathematics degree at Novosibirsk State University, Sergey told him in no uncertain terms that he could not continue to run the science station and the park alone and asked his son to join the business. Nikita accepted, albeit reluctantly at first, and they began their work cataloging the bones at Duvanny Yar. Sergey brooks no romantic nostalgia for the Soviet Union, and Nikita is even more disparaging. "The Soviet Union grew on the bones of tens of millions of people," he says. "I am pretty much sure that if I would be living a century ago, I would be long dead already." Yet the influence of earlier Soviet scientists such as Vernadsky endures in their work, and their experiences of perestroika and the hardships of the subsequent years have shaped their thinking and approaches toward ecological restoration. It is impossible to understand how the Pleistocene Park came into being without acknowledging the contingent histories of strife and brutality found locked within the permafrost soils of Chersky. The town was formed off the back of gulag labor, from the excavation of the earth to the mining of gold that filled the coffers of Moscow. How these difficult and, indeed, rawer pasts collide with the deep histories of Pleistocene skeletons is generative of the specific ideals of redemptive utopia found in both Sergey's and Nikita's thinking. The Soviet Union and the resulting fate of Chersky might also be described as ruins, and these shallower forms of decay inflect the narrative of survival and redemption found in the park's ethos. To find stability and safety, the Zimovs must dig deeper into the past, to a time of "natural" balance, self-regulating nutrient cycles, and predator–prey relationships. Replicating such a balanced world demands bypassing the fraught and tangled histories of extinction and failed political utopias and turning instead to the invisible and imagined deeper subterranean that might provide the solution. Sergey and Nikita are not so much "zealous masters" as stewards of a redemptive future.

(Re)emergent Landscapes and the Intricacies of Reproduction

Of course, the Pleistocene Park cannot work without the introduction, or perhaps "return," of Pleistocene mammals—or at least their modern-day equivalents—to the tundra. With the encroaching climate crisis and the degradation of the permafrost, the Zimovs are in a race against time, not only to transport the necessary animals to the park but also to ensure favorable conditions for their survival, adaptation, and procreation in an ecosystem not yet fit for purpose. The vision they have is one of a self-sustaining rewilding project, but the park's current state is far from that. Many of the animals die—Nikita reckons about half. His voice develops a sort of frustrated sadness as he recounts the loss of an entire herd of elk, explaining: "It's the cold. Not all of them can adapt very well." The failure to adapt is unsurprising, of course, as these creatures, brought from across the planet, are being expected to bridge not merely a geographical gap but also a temporal one; the prehistoric ecosystem the Zimovs are attempting to re-create has lain in ruins for centuries, and these animals have no parents to teach them how to live in such a harsh climate.[38] The Zimovs are also working without a blueprint for their project. The learning curve to begin laying the groundwork of restoration has been steep, from Sergey finding out the hard way that fences are necessary to keep horses from running away to the need for veterinary care and specialized medicine to fight various parasitic infestations to the necessity of providing food through the harsh winters for creatures that have not yet learned to forage in several meters of snow. And all this is assuming the animals' successful transportation to the park, something that is nowhere near a given.

The bison problem takes up most of Sergey and Nikita's time during my stay and is the subject of several overheated arguments, neither man knowing quite what to do about it. The park already has a single male bison, a European wisent purchased from an animal park in Russia, but he is aggressive and needs the company of a viable mate to be useful, so the Zimovs have spent the past few seasons trying to procure more bison. A year earlier, they had run a second crowdfunding campaign, after their first appeal to pay for a herd of yaks had raised more than $100,000. This time, they wanted American bison—larger beasts than their European counterparts and better for the ecosystem, as they would act as similarly destructive substitutes for the mammoths, bulldozing trees and churning up soil. Sergey and Nikita had sourced a herd of twelve baby bison

Bison are key to the Pleistocene Park's plan to have heavy, destructive mammals churning up soil and destroying trees. For years the park was home to a single wisent (a European bison), but with the arrival of a herd of American bison from Denmark (pictured in the background behind the wisent) breeding is now a possibility. Photograph by Nikita Zimov.

in Alaska, titled their crowdfunder "Bison to Save the World," and crossed their fingers. This time, donations limped over the two-thirds mark, the logistical bureaucracy hit an impasse, they were unable to find a pilot for their specially sourced ancient DC-4 aircraft, and every day the bison remained in their Alaskan pen was costing more and more money. Nikita recounts:

> We still didn't get our bison, which is also another big complication in our life. I am not sure we will be able to take those Alaskan bison. I don't know how much money we already lost. Like . . . twenty to thirty thousand bucks. So if we will now cancel it will be just lost money. . . . We try to push, push, and I don't know. We will wait a couple more weeks and if we don't set the actual date when they are coming and we don't have a contract with the air company I think we will cancel that. Somehow we blew ninety thousand euro on another twenty bison from Europe in 2014.

> My dad managed to find some very weird organization . . . we paid for those bison to some foundation, the foundation told us that some farmers own those bison and then we paid the money for some transportation company to transport animals and I think at some point Russia kind of closed borders for some disease and I think this company . . . what's it called when they steal money and trick people? Eventually we paid for the bison and money never came back and we paid for the transportation and money never came back so it was a disaster.

The logistics involved in transporting animals—whether by truck, ancient aircraft, or leaky boat—are vastly complex and extremely expensive. Every loss of a creature is a setback in the effort to establish a working population, and even the slightest thing going wrong can mean the hemorrhage of huge sums of money. Nikita confirms that every bit of excess profit made by the research station is funneled into the park and toward the purchase of new animals. Crowdfunders and donations are the paramount sources of funding, and the Zimovs work to entice a steady stream of film crews and reporters to visit, harnessing their interest to garner as much media attention as possible. In what is clearly a painful decision, Nikita eventually pulls the plug on the Alaskan bison deal and takes a huge financial hit. The next summer, he and Sergey undertake perhaps their most difficult journey yet: using a specially designed truck, they set out to transport to the park twelve baby bison bought from a farm in Denmark. A GPS tracker on the truck makes it possible for people to follow their journey online in real time, watching as they travel roads that become more and more perilous, segueing from asphalt to dirt to nonexistent. From that point the truck is transported by river on a huge barge, then a smaller barge, and then finally is winched onto land at the Chersky shipping port. On their arrival at the park, the bison are to begin their new life in a strange new landscape, tasked with surviving, breeding, and "saving the world."[39]

When it comes to the animals that do make it to the park, there is no guarantee they will survive, displaced as they are from their homes, which are often thousands of miles away with completely different climatic conditions. Many of them struggle, even with

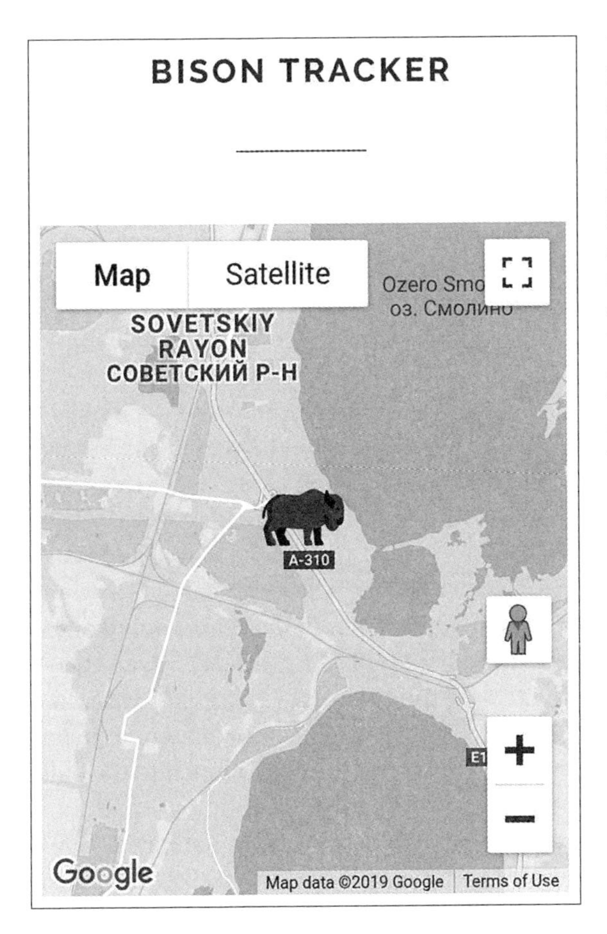

The Pleistocene Park successfully sourced a herd of bison from Denmark in 2019, and Sergey and Nikita drove the animals almost the entire way in a specially designed truck. A GPS tracker attached to the truck allowed interested observers to follow along online during the several weeks of the drive.

the help of the park's rangers. On one of my visits, I meet a blind yak, huddled forlornly in a pen. Nikita does not know why she went blind, although he is hopeful of a successful outcome as she has caught the romantic attention of the bull yak: "This male doesn't really like other girls but every time he sees that blind one he tries to take care of her and is trying to mate with her." He tells a ludicrous tale of undertaking a perilous journey by boat to Wrangel Island to collect a herd of six baby musk oxen, only to discover that a polar bear had broken into the compound and eaten one of the creatures, scattering the rest of the herd. Nikita and his team spent ten days hunting for another herd, an effort that ultimately ended in failure:

"We caught one male, then another, then another. All males! We got them there and we have three roaming somewhere in the park and they can . . . well, they learned how to pass through the fences. So we risked our life, we made this complicated expedition, but we didn't establish the population." These reproductive failures frustrate the redemptive arc of birth and restoration, while the expectation that animals will provide particular ecological or biological services is transgressed regularly. At the time of this writing, Nikita's second expedition to Wrangel Island, this time to find some female musk oxen, has been thwarted by bad weather and postponed.

The creatures of the Pleistocene Park enter into a sort of biological contract, in that it is their task to make the park a success by reproducing and thus helping to create a working ecosystem. The males are configured as virile studs and are allowed to roam where they please, and while the name of the park's lone European bison, Mussolini, is a darkly comedic reference to how violent he is around other creatures, there remain undertones of masculine dominance of both sexual and eco-spatial kinds. The females, of which there are bigger numbers to increase the chances of successful pregnancies, are breeding machines. There is no room for sentiment here; the blind yak receives care because of the sexual attraction the bull has shown toward her, not from any sense of compassion. Nikita and Sergey both freely admit they are not conservationists from a species revival perspective: "I personally don't really care about the animals that much," Nikita tells me. "We're trying to create the whole system. . . . I'm not an animal rights activist; I'm much more pragmatic and practical." The importance of the animals brought to the park is not determined by whether they can thrive and reproduce and thus serve as indicators of species health, but rather by whether they are able to provide the labor necessary to re-create the ecosystem of a deep past.

The entire process (usually headed by Nikita) of locating suitable animals, finding willing sellers, raising the money for the creatures themselves and for the transportation, and then undertaking the perils of the journey, followed by keeping the animals alive and enclosed within the park's fences while encouraging healthy reproduction, is a complex balancing act beset with difficulties, both for the animals and for the humans involved. One of the fundamental tenets of rewilding is that a "lighter touch" on the part of humanity will allow natural processes to produce authentic ecosystems without intervention; such an approach is far from the reality at the park.[40] There is undeniable—and masculine—hubris in attempting to restore an already extinct ecosystem, no matter how much the Zimovs refuse to see themselves as "zealous masters." That the creatures are mobilized as parts in a reproductive assembly line serves to further categorize the park as a techno-utopic space within a narrative of ecological restoration and circulation. Yet when faced with the ruined Pleistocene ecosystem, the fantasy of circulation and legacy breaks down. In a sense, the bones beneath the animals' feet are configured as their ancestry—a knitting together of broken timelines and a curation of a strange sort of legacy: a discontinuous one, with centuries of learned behavior passed down through generations completely missing. I find it difficult to imagine a thriving population of yaks when I watch the blind female tentatively scraping at the alien tundra floor. How will she know, if indeed she does get pregnant, how to teach her young to survive here?

Animal Agencies and Skeletal Afterlives

It is a strange thing to imagine a newly released baby bison from Denmark, likely shaky-legged and disoriented after a traumatic journey, rooting around in the park's soil and coming across a Pleistocene bison bone. This is a merging of past and present with a view toward redemptive futures, a meeting of life and death, fractured and fragmented through a permafrost landscape that is in flux. What exactly happens in this jagged space when living creatures are harnessed to do the work of their long-dead "relatives," out of time, out of place, out of joint, burdened by species affiliation?[41] Their lived reality involves fences, bales of hay, and shipped-in feed, medicines, and vaccines, along with constant monitoring by the park rangers. The knife-edge of survival is made keener as the permafrost thaws and retreats farther, and the remains of the past become more visible; the animals' task grows weightier as they struggle toward the future imagined for them by their human handlers. This is not a future of their choosing, and they often subvert it: refusing to mate, disappearing into the larch forest, shirking expectations.[42] Then there are the instances when creatures become tangled in other disruptions, through the high death rates of bitter winters, through sexual incompatibility, through overblown bureaucracy and logistical nightmares. Nastya tells me a shocking story of another attempt to bring some bison to the park a few years back: the creatures had made it all the way to Yakutsk on a livestock plane, only to be commandeered at customs and sent to the governor of Sakha's personal animal park. The Zimovs know the bison are there, but they have no way of retrieving them or lodging a criminal complaint; if the collapse of the Soviet Union has left one lasting legacy, it is that of institutional corruption.

The displaced creatures that are transported successfully stumble out of trucks and onto the land; as they begin to explore their new home on top of permafrost, rooting for food, their hooves sink down into mud. There is a sense of transgression here, as they con-

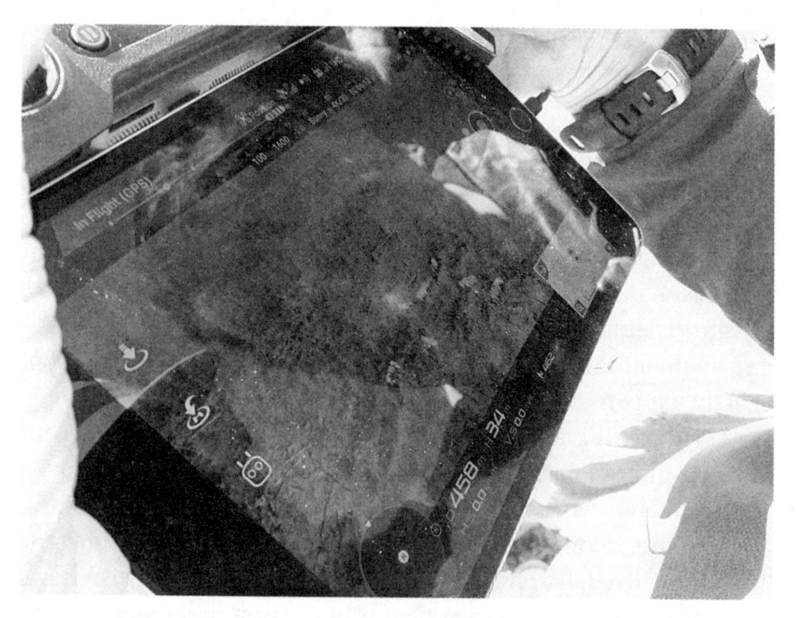

The Pleistocene Park is not huge, but much of it is inaccessible or difficult to reach because of the terrain. Animals can disappear for weeks. Nikita is pleased when drone footage reveals his herd of horses, with two new foals.

verge on a small area of tundra from places across the globe, but also as their bodies transgress the boundary between aboveground and belowground. Thawing permafrost facilitates this meeting of bodies and temporalities, introducing the animals to an imagined ancestry they have no knowledge of and a utopian future they have little hope of providing.[43] I am struck by this as I watch drone footage of a group of wild horses traversing the spindly larch forest; the horses are not physically reachable because they are too far into the undergrowth. Nikita has not known their whereabouts for weeks, so he is grateful to catch a glimpse, and even happier when he notices there are two foals among them. That animals breed without help from rangers is vital to the success of the park, but it is often a difficult process, and certainly not a guaranteed one. These animals have their own legacies, their own agency, whether they breed and produce young or not. The park is fenced, but the area they can roam is relatively large, and much of it is inaccessible by humans—weeks might go by without a ranger seeing a herd or a flock. Disappearing into the forest, giving birth on their own, refusing to breed—such acts disrupt the utopic fantasy of the park. These legacies are not linear, nor are they continuous. A lineage may stop with infertility or even death, the fleshy corpse sinking into the soil to join the myriad bones that already reside there and become part of an ongoing ruination of utopic futures—a legacy of refusal rather than of reproductive success.

The meeting of bones and bodies, the imagined utopia and the ecological ruins, the deep past and the shallow present—a fraught intercourse brought together and held together uneasily by the subterranean permafrost landscape of the park. In particular, the bones that remain press upon their living counterparts in ways that suggest they have afterlives that linger thousands of years later, pushing back against the finality of extinction, but also disrupting the neat linearity that the Pleistocene Park envisages.[44] According to Jennifer Wenzel, afterlives are "unfinished business, incomplete projects . . . broader questions of time, narrative, and nation: how the past is put to use in the present."[45] The unfinished business of the skeletal afterlives in the park lays bare the ruined ecosystem experienced by the rewilded animals; their refusal or inability to *become* the thriving Pleistocene landscape reveals frustrating discontinuities that only further highlight the temporal chasm between bones and beasts. To understand what the bones of Duvanny Yar mean is to understand how extinction is a process of intergenerational unraveling. Ecosystems and thriving animal populations are built up gradually, over time, animal bodies slowly becoming attuned to the environment, young growing into their milieu by way of learning from elders. It is impossible for the rewilded creatures to take up the mantle of their so-called ancestors; these bones cannot teach, these bones cannot guide. They produce ruptures in the assumed linearity of time—material reminders of an absence or an emptiness.[46] Their afterlives speak to an ecosystem that no longer exists and can never exist again, buttressed by the ebb of permafrost.

This does not mean the Zimovs intend to stop trying, despite the many setbacks they have encountered. They are, of course, as aware of the park's failings as they are of its potential, and this is the reason they feel the reverberations of the past so keenly. Gastón Gordillo calls this a "voiding of space," in which the emptiness of the present wrought by a vibrant past presses on the future.[47] In particular, this is an imagined utopia in the vein of Vernadsky that might struggle against the crushing weight of a coming apocalypse, a sort of Noah's Ark that will protect valued life from the flood. But to create this utopia necessitates a dystopic *outside,* a fence-ringed haven of safety within which animals can fulfill their given task of saving the world. This is, of course, a fantasy. Notwithstanding the physical breaches of the park's fences by unruly creatures and hungry locals, the ability to create a utopic boundary that nullifies future catastrophe rests on the capacity to harness the deep past for a redemptive future. The failures and absences of the current configuration of the park confound this vision; the living, breathing, trampling animal bodies above the ground are supposed to rebury their ancestors, compacting strata, forging onward through time and reversing ruination. What they actually do is often the opposite. Their legacies may be to join the piles of bones in the ground, gathering afterlives of their own that speak to futures that could have been. Wenzel imagines afterlives that disrupt the considered order of things, how the failures and missed chances of the past offer opportunities of resistance in the present.[48] The Pleistocene Park is almost a reversal of this: the imagined nostalgia of a working ecosystem spotlights the inability of the rewilded creatures to produce planetary redemption. What happens to time here? Instead of following a utopic linear progression, as strata gradually layer and layer once more, the failure of the park and its creatures to stay the course leaves behind a sort of temporal stickiness, in which the uncanniness found between life and death haunts across time, revealing the terrifying prospect that there might be no redemptive future at all.

The Subterranean as a Future Repository

When imagining Duvanny Yar thousands of years from now, we might ask what the archaeologists of the future (human, more-than-human, alien) will find there. Will there be a trace of a brazen idea, a global redemptive project born from a prehistoric graveyard and the gathering of creatures from across the world? Will there be a tangible Zimov legacy in the frozen ground? For all their posturing and self-promotion in their media appearances, the Zimovs realize that their project will not "save the world" in the sense of maintaining the status quo of unfettered environmental destruction; rather, they hope that their work will bring about a shift in behavior inspired by the prehistoric past they wish to replicate. As Nikita tells me:

> I'm not very optimistic about future of the world, and what we're trying to do. . . . If we will make these ecosystems it would be, sort of, in the future it could be basis for, how you say, more efficient sustainable living. So there is that. That's pretty much the philosophical idea of the park. So it's not only for science or climate . . . climate change, this is just part of the story, but I think these ecosystems can be beneficial. Artificial ecosystems are not very effective, there is various reasons for that, but if we try to develop this more efficient land use based on what nature teaches us that may be good in the future.

Nikita's admission that he is not optimistic about the future comes not from any failure to trust his project but rather from his fear that he will not have the time or ability to enact his vision throughout the world (or at least throughout Siberia). The remains of the Pleistocene ecosystem haunt him and Sergey across time, while any successes the park enjoys are frequently confounded by things going wrong; as Sergey once wryly told me, it is two steps forward, three steps back. None of these fears, frustrations, and doubts make it into the glossy media outputs or merchandise promoted by the park; instead, the Zimovs swallow any misgivings as they promote the dream that there once again might be a balanced, self-sustaining ecosystem. The permafrost, however, has other ideas.

To borrow one of the many phrases associated with the Anthropocene, permafrost thaw refutes the idea of a stable, safe operating space.[49] If the permafrost is to act as a deposit box for unstable and unsafe futures, the vast deathscape of skeletal remains at Duvanny Yar and elsewhere must be overlain with new strata. As much as the Pleistocene past is harnessed as an ecological utopia, the specter of Pleistocene overkill cannot be ignored, and the fact that the mammoth steppe ecosystem lies in ruins serves as a reminder that humans may repeat the process in the future. To maintain the subterranean imaginary as a space of unchanging stasis and solidity, with little of the dynamism that generates surprise and fear, this past utopia must be constructed once more *without* the presence of the skeletal remains of extinction. The physical presence of bones betrays the reality that the labor being done by the rewilded animals is failing to work—or at least is not working quickly enough to have much effect on the permafrost. The problem with imagining a utopic future ecosystem in which humans are able to live sustainably is that it ignores any attempt to live more sustainably *now.*

Returning to ruination as a "virulent verb," it becomes apparent how the thawing of permafrost and its forces of decay are ruinous in themselves; they allow for ruination to take place, troubling the boundaries of subjectivity by helping interrogate exactly *what* is doing the ruining and what is becoming ruined. The Anthropocene project is necessarily predicated on what becomes fossilized—our plastics, our toxic pollution, our radioactive dust—but what about that which is not fossilized and instead becomes ruined remains, or vanishes altogether? The process of ruination is not a one-way street, where all lifeless and crumbling wastes are produced by the destructive hand of humanity; rather, ruination is an emergent and surprising force of more-than-human agency.[50] The bones and bodies of long-dead—and extinct—creatures might point to the terrible power of anthropogenic destruction, but these remains are also material and temporal ruptures that force humans to face their own extinction through a particular Anthropocene narrative of precarious mastery.[51] The imagined utopia of the Pleistocene Park has no truck with this—there would be no point in admitting such a possibility. Instead, the Zimovs attempt to discard these terrible futures and look to deeper and more solid pasts in which the permafrost endures and the land becomes balanced once more through the efforts of both their heroic temporal tinkering and the reproductive abundance of their animals.

Extinction presses upon the land here. When speaking of legacy and the afterlives of the Pleistocene bones, it is important to also point out the legacies that the rewilded animals and the Zimovs themselves may leave in their wake. These legacies are intricately entangled; the Zimovs cannot leave their legacy of a fully working ecosystem without the animals fulfilling their expected roles—breeding, raising young successfully, embodying the behaviors of their so-called ancestors—but an expectation of continuity between a deep past and a shallow present cannot be predicated on a permafrost landscape that is constantly changing. The vertical progression of time and knowledge is easily eroded in such a space. Extinction is the dismantling of generational knowledges across time, but it is also the space in which new knowledges can emerge.[52] No, the animals at the park may never join up the temporal chasm of ecological extinction, but perhaps they might become something else; perhaps they might curate new ways of living with permafrost that do not correspond with a "return" of a deep past. This is in no sense a "de-extinction" (a topic explored in the next chapter); rather, it is a way to conceptualize extinction as a transitive force—a process of undoing and *re*doing discontinuously—that makes space for new forms of life among the ruins.

To end, I want to point to the possibilities and opportunities these remains present—by acknowledging that ruination might not always be a bad thing, by recognizing the agency of more-than-human actors and forces that *also* ruin, these alternative ways of thinking might begin to dismantle and disentangle the anthropocentric dialectic of control and imagined utopia. The Pleistocene Park does not exist in a vacuum; it is a product of thousands of years of deep histories and shallow histories, stratified, decayed, and entangled through the ruining forces of permafrost thaw. These histories might be imagined or forgotten, covered in layered dirt or washed away by rivers, nudged around by furry snouts or stricken from the textbooks. It matters how these discontinuous histories combine or come apart. What ruination can do is expose that which has long been obscured, stripping bare the layer upon layer of colonization, extraction, and slaughter that can be found in the soils of Siberia and beyond.[53] It can also help to ruin the seemingly endless reproduction of generational linearity and orientation toward the future. Coming to terms with this is difficult, but it is also essential for finding ways of responding to remains that are not predicated on prolonging the destructive practices of the Anthropocene and curating space for new future imaginaries that might allow us to live better with the remains of our virulence, but also the virulence of others.

The Lives Coming Back

One sunny afternoon, finding myself with little to do at the station, I wander down into Chersky. On the way, I pass ruined buildings, cracked and bowed by permafrost thaw, scavenged of all their useful materials. The town betrays fragments of its Soviet past: a gleaming Tupolev aircraft as a monument to the might of Russian aviation, a gigantic mural on the side of a building depicting cosmonaut Yuri Gagarin. A little farther out of town, I come across the cemetery. It is a small, ramshackle place; headstones are piecemeal, made from wood or repurposed materials (one is fashioned from the propeller blade of an aircraft). Some are scattered with dead flowers and faded ribbons, and many have fallen over or teeter dangerously. The ground is mushy and humped, and I am very careful as I navigate my way across the fetid permafrost, trying not to step on the mounds of raised earth. There is nobody around—nobody alive, anyway. I think

The Chersky cemetery: even the dead cannot rest soundly in a permafrost landscape that is thawing.

about the people buried here, sunken into permafrost, at the mercy of thawing soils that threaten to expose their bodies in death.

A few days later at Duvanny Yar, I ask Nikita if he cares about leaving a legacy, if it matters to him if his name or his work is remembered after his death. He considers the question in silence for a moment before answering:

> So there was a guy here, he was considered head of local mafia; for all intents and purposes he was a business manager, but not all his business was entirely legal, let's say. This was in the 1990s, so he was some sort of badass criminal who . . . well, most organized crime are trying to get money from you, you know, there is a man and he come to your business and say, "Okay, you pay me or I will kill you." He didn't do that, but he was doing some illegal business with diesel fuel. He also did lots of fish and truck driving, so he was like a little local corporation. He was in the mammoth

> business, definitely. He was trying to make money off of everything that was possible. But people actually loved him, so he was a good guy . . . and he was forty-nine when he died from cancer. And it was the biggest funeral in town, so there was two hundred people probably, and there was like forty-four cars in the cortege and two helicopters from the other regions brought people, brought a priest from the other place. It was about five or six years ago. And now there is nothing left from him. His family moved out to the mainland, and all his business that he was working on, creating this little Northern empire, that's all gone. Nothing left. His fishing camps are rotten, his trucks have been broken or taken away. And now people don't mention him too frequently already. Apparently, the only thing he left after he died will be his kids. So, it's the same here. In the eyes of many people, I live quite a cool lifestyle and doing important work, but if now on the way back we will flip the boat and I will drown, the only thing that will remain of me after twenty years will be my kids.

Sergey's legacy is undeniable; he is a famous and respected scientist, and the Pleistocene Park and the research station are his achievements. For Nikita, things are a little more complicated. He is aware of the weight on his shoulders, not just of carrying on his father's work but also of the future he wants to provide for his children. He and Nastya have three daughters, and I ask him if they will take over the running of the park from him one day, just as he had done from Sergey. His answer is evasive: "Let's just say, if by the time I have to be retired and someone has to do my job, it means I failed." Nikita is reluctant to allow his daughters to take on the dirty, exhausting work and long hours that running the park requires, but his words betray a more ominous fear—that if the park is not a success by the time he must retire, then it will be too late. I press further and ask him if he thinks it is ethical to bring children into a world he clearly has little hope for, despite all his brash claims of saving the planet. He explains: "I try to make it a better world. I am sure I will be able to teach my kids in the future to be environmental and spread the word. The more people like me, the more future generations will be like me. Maybe that's too proud to say . . ." Nikita sees his legacy much as he views the role of the animals in the park—he is teaching his young how to produce a better world. When the Zimovs refer to their plan as "the world's best," they are also constructing themselves as the world's best practitioners of such a plan—"Earthmasters" and custodians of a global legacy.[54]

This particularly heteropatriarchal imaginary of the park is possible only because the park is cut off, bounded by fences and geographical isolation; in a sense, it is Vernadsky's noösphere made flesh (albeit on a much smaller scale). But Nikita's insistence that he "cannot make an island here" rings all the more true when the transgressions of this island boundary are acknowledged: locals eating an errant ox, the logistical impossibilities of bison transportation, the refusal of permafrost to stay frozen. Utopia can be projected onto these voided spaces, but the Pleistocene Park is also inseparable from its global intentions and practices.[55] Ruination through permafrost thaw occurs at both global and local scales, the decay of strata forcing a temporal shift that reveals the *potential* to create a utopia but at the same time takes it away.[56] The remains of prehistoric animal bones are ruptures that galvanize action and point to the possibility of failure, while the struggles of the animals to adapt and fulfill their ecological role as saviors cause further anxiety that the projected redemption might never come to pass. These creatures, both dead and alive, both vastly ancient and awkwardly present, generate their own afterlives that unsettle the vertical stacking of time, eroded by the discontinuity of permafrost.

What is actually happening in the park is that permafrost thaw is opening up an awkward temporal chasm to reveal the ruins of an extinct ecosystem. This is a chance to redefine extinction as a generative process, something that does not end completely and instead manifests itself through its traces found in the present. By accepting that ruination is an ongoing process and something not necessarily done by humans, and by acknowledging that ruination collapses both pasts and futures, might we enable a more nuanced (and certainly not utopic) interpretation of the Pleistocene Park and its permafrost? Rather than viewing permafrost as an anthropocentric safe repository for the coming apocalypse, perhaps we should understand it as an ephemeral and surprising materiality that provides

At the summer solstice festival in Chersky, the entire town gathers to celebrate the "greeting of the sun" with a barbecue, dancing, and games. Nikita had always won the disk-throwing game, but this year he is beaten by a local man.

the conditions for futures that may not include us. Isabel Stengers writes: "Ruins may also be alive with partial connections, connections that do not sustain great entrepreneurial perspective but demand a capacity to learn from and learn with, and to care for what has been learned from."[57] Allowing for this ruination and accepting the unsettling legacies of these temporal and material remains might, in fact, give us tools *not* to save the world, but rather to build better from the ruins.

A few days after our conversation on the banks of Duvanny Yar, the whole town attends the Ysyakh celebrations around the carved Sakhan *sergeh* poles on the outskirts of town. It is a beautiful day, and there is a barbecue, a singing competition (won by Dasha, Nikita's daughter), and a variety of games. As much as I feel removed from this annual tradition, I also feel a sense of belonging: I am part of the research station, albeit briefly, and the locals are used to all manner of Gore-Tex-clad foreign scientists showing up on their doorstep.

The work at the station and the park are as much a part of Chersky as the old Tupolev and the Gagarin mural, despite the occasional eaten animal and other bumps in the road. Yes, the Zimovs are removed, and remove themselves, from the specific milieu of Chersky; they were not born here, and they leave every September for more hospitable climates. Their work with the park fences them off—both literally and metaphorically—from the embedded histories of the town, from its terrible beginnings as a gulag to its current state of economic and climatic hardship. But today, at the greeting of the sun, everyone is welcomed. People chatter animatedly in Sakhan and Russian. Old women in traditional Sakhan dress spontaneously dance together. It seems that nobody will be able to lasso the reindeer antlers, until a wizened old Evenki finally does, to great cheering. And down in the sandy wrestling pit, a local man manages to throw the heavy disk farther than anybody else, ruining Nikita's unbeaten run and thus ending one legacy and beginning another.

4
Blood

The first proper winter blizzard comes to Yakutsk about halfway through my stay, with snowflakes so large and thick that visibility is almost nil; the snow whips horizontally, viciously stinging any exposed parts of the body. Within about three hours, the light film of compacted slippery ice on the ground is blanketed in a thick layer, and the November sky is dark at 3:00 p.m. It is hazardous outside—Yakutsk is poorly lit and full of potholes—and inside every building a trail of gray slush leads from the outer door into the interior. Luckily, I am staying a mere five-minute walk from the Mammoth Museum, so I arrive there relatively unscathed. I have been put in touch with a tusk hunter called Sasha who is employed by the museum to find mammoth carcasses, and he has invited me to visit the laboratories. He had found an ancient horse a few months earlier, and Russia's top prehistoric zoologist is flying in from St. Petersburg to assess the body. When I arrive, snow-soaked feet encased in blue plastic shoe covers, I am treated to celebratory tea and cake and introduced to a team of South Korean geneticists who have just arrived from Seoul. I figure out very quickly who I am dealing with. The head of the delegation is Hwang Woo-Suk, a biotechnology expert and veterinarian who made global headlines in 2005 when he claimed to have cloned human embryonic cells—a world first. While the scientific community was grappling with the ethical implications of this, allegations emerged that not only had Hwang pressured his female lab assistants into donating their eggs to his experiment, but he had also falsified the results. Overnight, he went from being dubbed "the pride of Korea" to being blacklisted and disgraced, barred from conducting any governmental stem cell research, his state funding withdrawn.[1]

But Hwang, despite being under criminal investigation, was not done with embryo research. He retreated into the private sector and used his veterinary expertise to set up a pet cloning company, charging wealthy patrons—Barbra Streisand is a client—hundreds of thousands of dollars to ensure that their precious fur babies will never "die." There was nothing the scientific community's ethical and rigorous peer-review process could do about it. A few years later, Hwang began showing up in Yakutsk. A whole section of the Mammoth Museum is dedicated to him and his team, valorizing the great and pioneering science done by Hwang's laboratory in Seoul, Sooam Biotech, and the opportunities for collaboration with Yakutsk's North-Eastern Federal University. Hwang's personal fortune, amassed from his pet cloning business, financed the building of a state-of-the-art laboratory at NEFU—something the university could never have afforded on its own, given the pitiful funding it receives from the Russian Academy of Sciences. In exchange, Hwang requested access to any well-preserved specimens of prehistoric creatures found in the permafrost. He wanted to take samples back to his lab in Seoul, to play around with and experiment on ancient DNA, a much greater challenge than dealing with living cells. To put it bluntly, Hwang wants to clone a mammoth.

After the introductions, I listen as Hwang, who does not speak Russian, converses in stilted English with the head of the laboratory, Dr. Semyon Grigoriev, about his requirements for the visit. He is clearly uncomfortable with my being present, likely assuming I am some sort of undercover journalist, but he has no good reason to ask me to leave. Instead, he smiles rather too broadly at me and answers my questions with a single word or a mere laugh while he attempts to align his schedule with Semyon's. His flight back to Seoul is at five the next morning, so he wants to spend some time alone with the horse a few hours earlier to take samples and put them immediately into an ice cooler to keep them frozen. From there he will dash to the airport for his flight, taking the cooler on board as hand luggage. It is unclear how legal all this is. Nonetheless, it is clear that I am not invited to the 3:00 a.m. genetic harvesting soiree, so my only opportunity to view the horse is now. I follow Hwang as we are ushered into the laboratory where the specimen is waiting.

The horse has been thawing for a while now so that its flesh will

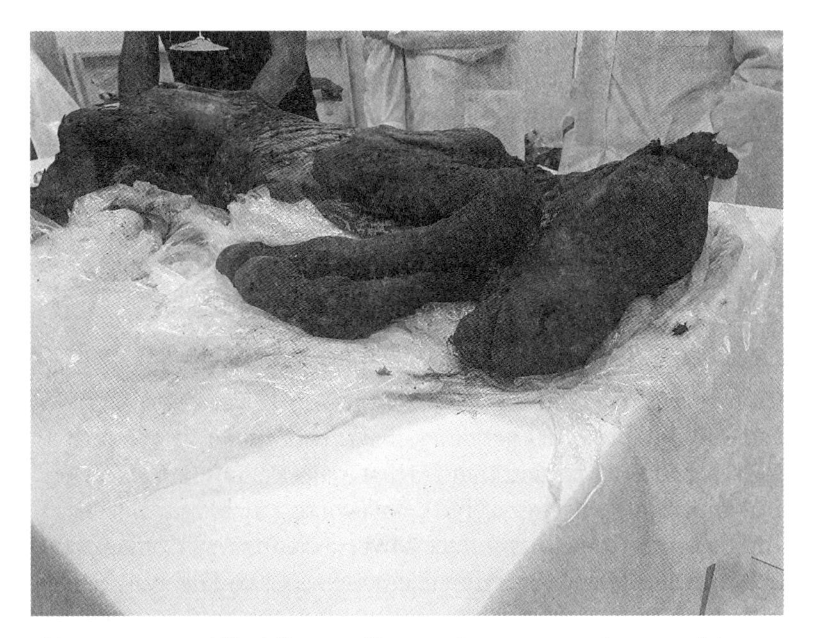

This 42,000-year-old foal discovered in permafrost was so well preserved that liquid urine was found by the team at the Mammoth Museum.

be soft enough to allow samples to be taken. The whole room stinks; unsurprisingly, prehistoric horse corpses don't smell too great. According to the results of the radiocarbon dating procedure, the creature is officially around 42,000 years old; it is estimated to have been between ten and twenty days old at the time of its death. It is remarkably well preserved, with no pieces missing, and every bit recognizable as a horse. The zoologist circles the body, poking, slicing, lifting bits up, and explaining as he goes. He does this in Russian, so I am denied Hwang's thoughts on the matter. We retreat back into the office for more tea and cake, and the conversation turns to dinner plans, which again obviously do not include me. I awkwardly make my excuses, and within minutes I am back out in the blizzard and making for home.

While Hwang is keen to get his hands on any prehistoric genetic material he can, it is the mammoth that is the prize. He is adamant that there is a perfectly preserved piece of DNA (a single nucleus, to be exact) out there that will allow him to practice his cloning technique—DNA contained within a cluster of cells that was frozen so quickly that no degradation whatsoever has occurred. The chances of finding such a specimen are extremely slim; every mammoth found in the permafrost so far has yielded only degraded DNA, useless for cloning. But there is another technique for mammoth de-extinction, the one currently being pursued by George Church's laboratory at Harvard University, which involves a sort of cutting and pasting of genes using the genetic engineering tool called CRISPR. Once an extinct genome is sequenced—the mammoth's genome was sequenced in 2015—CRISPR can be used to isolate specific genomic features found in sections of DNA, which can then be swapped with and plugged into other genomes, like that of the Indian elephant.[2] By "switching on" certain features that make a mammoth, as it were, and implanting the engineered embryo into a viable surrogate, scientists believe they can produce a hybrid elephant with shaggy fur. This method of de-extinction is much likelier than Hwang's to be successful, but, of course, it raises the question: Would an elephant with cold-adapted blood and smaller ears *be* a mammoth, or would it be merely an aberration? And could the creature successfully breed with other hybrids, or would it be destined to become a sort of reverse endling, both the last and the first of its species?

George Church visited the Pleistocene Park a few months after I did, accompanying Stewart Brand as part of the filming of a documentary about Brand's life, but also to scope out a potential place to put his future mammoths.[3] A mammoth would require a mammoth habitat, a cold and vast territory with high-productivity grassland, and where better than the Pleistocene Park, a moniker consciously reminiscent of the most famous example of fictional de-extinction? Mammoth de-extinction, however, is low on the list of priorities for the park. I ask Nikita about his thoughts on the subject, and he explains that they walk a fine line; while the obvious shock value of de-extinction could gain the park recognition and funding, placating naysayers and deflecting criticism would come along with that. "You have a lot of people believing in God," he says. "And they don't like this mammoth return. So I try and use it to bring attention to the park, but I don't want any of the criticism!" The official line is: you can use our land for a habitat, but leave us out of the actual process itself. Meanwhile, the rumblings in the scientific community and the science communication media suggest that mammoth de-extinction, by whatever method, is not too far away.[4] How, then, does this new scientific approach renegotiate ethical, political, and material boundaries as the category of life is troubled, sliced open, and forced apart by new forms of biotechnology and cryotechnology? How does extinction become something to be controlled and manipulated through the laboratory, and what happens if the Pleistocene Park once again hosts the mammoth, tasked with a mission of survival and redemption?

A few months after I leave the frigid temperatures of the Yakutsk winter and return to London, I notice that Hwang has made the news once again. The 3:00 a.m. sample collection undertaken during my stay has been unsuccessful in producing any viable cells for cloning. Hwang had returned to Yakutsk and performed another autopsy on the creature in an attempt to find better-preserved bits of flesh, slicing the body fully open and extracting organs and viscera, in the process finding both liquid blood and liquid urine. An article in the *Siberian Times* states that "researchers are confident of the success of this project" and goes on to report that a surrogate mother has already been selected to carry the cloned cells to term.[5] If the horse manages to live, it will be a world first, but such an "achievement" comes with many ethical questions.[6] Assuming the researchers' confidence pays off, it remains unclear what will be done with this cloned horse: Will it spend its life in a zoo or be put to work elsewhere? Will it be forced to reproduce, or is it destined to become another endling—a curiosity of its species out of time, surviving, but with no life at all? It is difficult to see the cloning of this creature as anything other than hubristic posturing by a disgraced man desperately trying to claw back his lost adulation.[7] When it comes to de-extinction, however, and the manipulation of ancient genetic material, there are deeper elements at work: a cryopolitical renegotiation of frozen life as a response to a heating planet, not least materially, as permafrost retreats and reveals hidden depths in which prehistoric carcasses have lain dormant for millennia and are now, in a way, awakening, acting, resurrecting, and reconfiguring what extinction means in the Anthropocene. No beast embodies this process more than the woolly mammoth.

Mammoth Histories

The first—or at least the first documented—full frozen mammoth body was found at the mouth of the Lena River in 1799 by the Evenki hunter Ossip Shumakov.[8] Disregarding warnings from other Evenki not to disturb the carcass, Shumakov hacked off the tusks and sold them to a merchant in Yakutsk, who presumably spread the word around town about the find. Several years later, the story reached the ears of the botanist Mikhail Adams, who led an expedition to exhume the mammoth's body and bring it back to St. Petersburg. The expedition was mostly a failure, with Adams arriving in August 1807 to find a rotting mammoth corpse that had been torn apart by scavenging animals. The majority of the organs and trunk had been eaten completely, and he was able to retrieve only most of the head and two of the feet, which had remained encased in permafrost. However, Adams scored a big success with the collection of the skeleton, and once back in Yakutsk he was able to procure the creature's tusks from the original buyer. By consulting sketches of the complete carcass drawn by the merchant based on Shumakov's oral testimony, Adams and a colleague were able to assemble the first ever mammoth skeleton for display to the world. Although not without its errors—the tusks were placed the wrong way—this reconstructed beast became integral to the new paleontological understandings of prehistoric creatures. The fact that the Adams mammoth had been found with skin covered in hair proved that the creature was adapted to living in cold climates and was not simply an elephant that had wandered off course.[9] Along with the fossilized remains of mammoths, the Adams specimen and the subsequent discovery of other intact bodies in the permafrost contributed to an understanding of the mammoth's prehistoric habitat, behaviors, and eventual extinction—although debate still rages as to whether this extinction was brought about by Pleistocene overkill or climate change.[10]

With these discoveries, mammoths quickly became objects of

The Adams mammoth was the first mammoth skeleton ever assembled and displayed at the zoological museum in St. Petersburg, where it remains to this day. Photograph by Charlee Michelle Gothard.

curiosity, and displays proliferated in museums and exhibitions: skeletons, life-size models mocked up with hair, and even the preserved specimens themselves. The mummified corpse of a male baby mammoth, given the name Dima, was discovered in the Magadan region in 1977, and another baby was found on the Yamal Peninsula in 2007. The latter, a female called Lyuba, was in even better condition than Dima, so much so that she still had her mother's milk in her belly. Both Dima and Lyuba currently travel the world in temperature-controlled crates and are put on display in glass cases for the public to marvel at. Amy Lynn Fletcher refers to them as "boundary objects": museum specimens that are able to travel through time between epochs and act as physical manifestations of how the past is mobilized in the present and future.[11] The Melnikov Permafrost Institute's permafrost cave is home to a plaster cast replica of Dima, having been the original storage facility for the body until it was taken to St. Petersburg. Other mammoths have achieved similar fame, not least the Jarkov mammoth, which was the subject of a bizarre expedition led by French explorer Bernard Buigues. Armed with little more than a vague tip from a Dolgan boy and a hefty wad of cash, Buigues took a crew into the tundra with the intention of drilling the creature out of the frozen permafrost. The Discovery Channel documentary *Raising the Mammoth* includes faintly otherworldly footage of the beast, still mostly encased in a block of ice, being winched into the air by helicopter and flown over the snowy landscape to a permafrost cave, where it remains to this day.[12]

Today, mammoth finds are relatively common. Thawing permafrost and a surge in the numbers of tusk hunters plying their trade in

The mummified baby mammoth Dima in his glass case at the St. Petersburg zoological museum, and the plaster cast replica at the MPI.

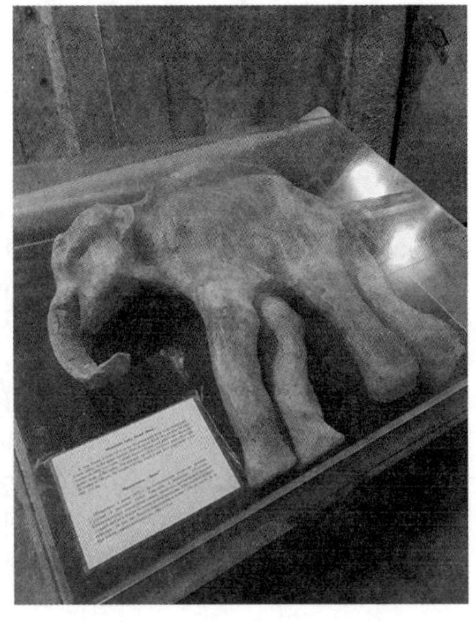

the summer months mean that carcasses in various states of preservation are discovered on a regular basis. Semyon Grigoriev made the news several years ago when, on an expedition to Northern Sakha to follow up on a mammoth discovery, he found the creature so well preserved (it was the first mammoth to produce liquid blood) that he promptly sliced off a piece of flesh and ate it. The Mammoth Museum has hosted researchers from all over the world who are eager to get up close and personal with prehistoric specimens; they have included George Church and Stewart Brand, as well as a documentary team that filmed an autopsy conducted on the mammoth whose flesh Semyon tasted, a female nicknamed Buttercup.[13] The documentary plays on a loop at the museum, near where visitors peruse the selection of bones, teeth, and skin on display, while the hacked-apart carcass of Buttercup now resides in the museum's walk-in freezer—just next door to where we met the horse—looking slightly worse for wear in a plastic bag. Hwang is intimately familiar with Buttercup; his team took several samples from the autopsy, including a vial of blood, in the hope of finding viable cells, but once again they met with failure.

The Siberian mammoth is now the focus of multiple vested interests. Having lain dormant in the relative peace of extinction, locked away for centuries in a permafrost tomb, she now finds herself the subject of excavations, blasted with water cannons, chopped into pieces, and broken down into genes, her tusks shipped off to China and her DNA sent across the world to biotechnology laboratories.[14] Her body is now worth less intact, with the knowledge gleaned from the Adams mammoth spent and the bodies of Dima and Lyuba now relegated to the status of mere curiosities. Instead, the mammoth's intrigue lies in her future potential, either as riches in the form of carved ivory ornaments or as viable genetic material for de-extinction. This renegotiation of the mammoth as a material and a sociocultural object across time can be correlated with both her scientific and her popular trajectories throughout history: the Adams mammoth appeared at a point when paleontologists and geologists were grappling with questions of earthly revolution and extinction; today, the mammoth's situatedness within the Anthropocene narrative means that she embodies both an engagement with a deep past

The sad final resting place of Buttercup, the first mammoth discovered to produce liquid blood by Semyon's team in 2013. Her autopsy, during which blood and tissue were harvested to further de-extinction research, was filmed by the British public television network Channel 4.

and the anxiety of an uncertain future. Buttercup encapsulates such a future. The fact that she lies almost forgotten in the Mammoth Museum's freezer is a result of her failure to produce the necessary pieces from which she might be rebuilt in technoscience's image. Yet when Buttercup's body was first discovered in 2013, she was deemed a scientific miracle. It all began with a drop of blood.

(Un)broken Lineages and Blood Ties

When Semyon and his team first found liquid mammoth blood oozing from Buttercup's leg, they could barely believe their eyes.[15] Much of it spilled onto the floor before Semyon quickly held a vial beneath the wound to catch the rest; he then chiseled into another part of Buttercup's flesh, only to be rewarded again. Liquid blood meant that this mammoth was better preserved than any other mammoth previously discovered; liquid blood meant that—at least potentially, *theoretically*—Buttercup could be brought back to life.

The discovery of Buttercup's blood represented something *more* than mere scientific potential, however. Blood resonates and vibrates with notions of kinship, family, lineage, and ancestry. It invokes sacrifice and redemption, is part of certain rituals and folklore. Bloodletting has been used in medicine for centuries, and blood sharing can be both a sexual taboo and a way of forming brotherly bonds. As Janet Carsten states, blood has an "unusual capacity for accruing layers of symbolic resonance."[16] Blood dries and desiccates upon death; it is the life that literally ebbs away as the heart stops and circulation slows, liquidity clotting and jellifying right there in the veins. But blood also endures after death, through relation, through children and parents, sisters and aunts. The term *bloodlines* might refer to the unbroken lineage of a certain familial unit, or it might be used to describe the pedigree of a desirable breeding specimen. Blood is at once unique to a creature and potentially fungible through blood transfusions and other forms of mingling. Blood can be warm with life or cold with vengeance. For the mammoth, extinct as she is, it was assumed her blood had drained away with the last of her life—that is, until Semyon Grigoriev's chisel struck a well.

That's the thing about blood: it is uncontainable.[17] Extinction, as it is traditionally understood, is a containment. The life of the species has been contained by the passage of time, with a cutoff point that implies the creature—in this case the mammoth—remains locked in the past. What the flowing blood does, however, is wrest the mammoth back from her position in deep time. Her body is dead and frozen, her species lost to history, but the presence of blood reveals something more, something *livelier*. To the scientists who found her body, and many others like her, Buttercup is both the first of her kind and part of a long lineage of other mammoths. Lineage suggests linearity: a gradual and quantitative accumulation of time through mammoth parenthood. Buttercup had parents, and her CT scan showed that she herself had given birth eight times; her gigantic tusks revealed that she had lost one of her offspring before it was weaned, but seven survived beyond that point.[18] Buttercup left a legacy, and her offspring likely went on to have young of their own.

But at the same time, Buttercup's incredible state of preservation—and her bloody existence on the cold slab of the autopsy table in 2014—means that she is out of sync with the rest of her line.

How does de-extinction figure with blood? Both Buttercup and the prehistoric foal described at the beginning of this chapter are candidates for de-extinction through Sooam Biotech's cloning operation, and both of them produced liquid blood. This blood might be seen to embody a sort of continuation, in which the kin-making bloody bonds of mammoth-hood have endured across time and the disruptive designation of extinction becomes questioned. Is it still extinction if blood and cellular material have remained intact and viable?[19] When life might be found in forty-thousand-year-old blood, the temporal chasm between extinction and *no longer* extinct closes. It is as if those ancestral ties have reawakened through the vial of Buttercup's blood, and her potential for being brought back to life restores order to a supposed linearity of species lineage. The blood in the vial begins as a shocking and jubilant discovery in the field, but the changing materiality of blood away from the regulating force of a beating heart is tricky for the scientist: without the use of an anticoagulant, liquid blood will clot and harden, or, worse, the cells will decompose and the DNA inside will be useless. It is only once the blood is back in the relative stasis of a frozen environment—in this case, the laboratory at NEFU—that the continuity of the mammoth species might be said to emerge. Frozen, cold, or anticoagulated blood can be manipulated, controlled, shipped to other laboratories for viability testing for de-extinction, the timeline contained along commodity chains.[20]

De-extinction purports to continue the future lineage of mammoth familial units, but this time there is a caveat: de-extincted mammoths must now work for a living.[21] The de-extincted mammoth emerging from the laboratory requires a renegotiation of blood ties, kinship, and temporality through the labor demanded by humans. The actual messy business of de-extinction is not a case of laboratory cultured cells and blood sharing, but rather—and this is of course theoretical—an unpredictable situation of environmental becomings and cross-species encounters. Rather than liquid mammoth blood maintaining a continuous lineage through laboratory control, the blood coursing through the warm body of the first de-extincted mammoth will be unstable, disruptive, and discontinuous. First, the mammoth's mother will likely be a different species altogether—the Indian elephant, as the closest taxonomic surrogate. Second, the kinship bonds the mammoth will make will not be with parents and siblings, but rather with displaced and disrupted creatures of other species, brought from far away, all embodying an ecological role discontinued by their prehistoric "ancestors."[22] There is no continuity to be found here. Just as the animals brought to the Pleistocene Park are faced with the bones of long-dead creatures of the same species, the mammoth's return could never be a fusing of broken timelines, or a continuation of familial linearity; that the mammoth becomes tasked with the labor of fixing a disrupted ecosystem is indicative of an attempt to turn *back* time—a temporal looping that relies on the past becoming the future and vice versa, but at the same time attempts to practice continuity through reproductive prospecting.

The mammoth therefore becomes stripped of any individual agency or identity through a process of molecularization. The backstories of mammoths like Dima and Lyuba stick in our minds because these mammoths are whole bodies. Details like Lyuba's stomach containing her mother's milk matter because they illustrate the familial histories and legacies of mammoth life in the Pleistocene. These mammoths have faces, often so well preserved that they look as if they are merely sleeping. They have recognizable features that align with the way mammoths have been imagined throughout history, portrayed in cave paintings, carved ornaments, and even Hollywood movies such as the animated *Ice Age* franchise. They generate emotional responses and remind people of the anxieties of extinction. The sense of uncanniness to be found in the viewing of preserved mammoth bodies is one major reason that well-preserved permafrost finds make global news.[23] Yes, they are dead, but their intact, skin-and-fur-clad bodies (or heads) emanate a sense of liveliness at odds with the great age of their corpses.

Conversely, becoming molecular renders the mammoth as severed from this attachment to her bodily form and liveliness. Breaking the mammoth body down into pieces of flesh and DNA renegotiates the scalar and temporal boundaries of life and death in species conservation and de-extinction. Here, the added dimension of freezing as preservation generates a sense of plasticity around the delineated forms of life and death, disruptive agency, and human control. The molecular mammoth moves away from the display case and into the freezer, the hacked-apart and bagged form of Buttercup barely registering now that she has served her purpose as a cache of genes and blood.[24] Her frozen tissue is cataloged and archived in laboratories and cryobanks worldwide, labeled with her taxonomy as an abstract cipher for mammoths as a species and a genome.[25] Freezing the molecular mammoth turns the biopolitical into the cryopolitical, while the mammoth's status as "extinct" in the realm of a de-extinction laboratory further complicates and unsettles preconceived notions of what life is, where it is located, and how it becomes mobilized in the face of a melting world. Through this, the mammoth becomes an atemporal body, removed from her historical context in the ancient permafrost and forced into the space of the cryobank, where her genes become commodities to be stored and swapped, used in attempts to renegotiate exactly what life is and where it might be found.

Entering the Cryobank

A few weeks after I return from Yakutsk, I make a visit to the Natural History Museum in London's South Kensington. I am met at the employee entrance by Jacquie Mackenzie-Dodds, head of the NHM's frozen collection. She leads me down into the very bowels of the museum, far away from the brightly colored exhibits, shiny glass cases, and throngs of visitors. After a good five minutes of descending nondescript staircases and walking down seemingly endless corridors, we arrive at the cryobank. There is an office and a couple of large rooms, one containing a laboratory of sorts, the other housing the freezers. A smaller room adorned with warnings and sensors contains three liquid nitrogen tanks. The place is empty, save for a couple of people working at their computers in the office, and there is a distinct sense of being cut off from the busy museum upstairs. Most NHM visitors would have no idea this is here, and there is very little mention of the cryobank on the museum's website; it is certainly not a part of the exhibit and display space. Jacquie is happy to give me a tour of the facilities and answer my questions, providing they stay within the guidelines approved by the museum's public relations department. I am not allowed to ask anything about possible uses of the cryobank's material for de-extinction, or, indeed, about any applied conservation measures at all; I must also avoid any queries touching on ethics or potential geopolitical tensions. The NHM's cryobank is for archiving and research purposes only, and the museum refuses to speculate on how the sorts of material stored there might be used by more practical conservation bodies. Or, as Jacquie says by way of explanation: "We're not living, we're dead. The deader the better down here!"[26]

To those involved in de-extinction research, Jacquie's statement presents a challenge. While the genetic material is not alive in any normative sense of the word, by keeping the DNA frozen, the cryobank preserves the *potential* of the genes to produce life. In her work on cellular experimentation in laboratories, Hannah Landecker identifies as a key turning point in genetic research the implementation of freezer storage, as scientists recognized that freezing could preserve biological material. The ability to keep cells in stasis through freezing and have them "return to life" once defrosted revolutionized thinking around biological time. Landecker uses the notion of plasticity to identify what she terms "the thread of life," the idea that cells outside the mortal individual body become continuously linked across boundaries of sex, species, and even time: "It is the new form of immortality built into scientific life—disembodied, distributed continuity."[27] While life is very much put on pause in the cryobank, the point of freezing is to maintain this thread of life. Whether it gets used for conservation or de-extinction further down the line is irrelevant—the thread of life continues regardless.

Not all freezers are created equal, however; nor, as it turns out, are specimens. Many of the specimens currently taking up space in the NHM cryobank (which has a capacity of two million) come from what the museum calls the "legacy collection"—genetic material gleaned from stuffed and collected creatures from the institution's displays, some of them centuries old. Much of the material in this collection is poorly labeled, confusingly arranged, and highly degraded; Jacquie must sift through all of it, decide which specimens are worth keeping, and then determine which freezing process will work best for each. The rest of the material that gets frozen comes from field collections. Sometimes these are donated, either by environmental organizations or by private collectors; sometimes NHM team members go into the field and collect specimens themselves. The freezer room houses freezers set at minus 20 degrees and minus 80 degrees Celsius, and the liquid nitrogen tanks keep their contents at a temperature of minus 196 degrees. The temperature of a freezer greatly affects the quality of preservation. Freezing slows the movement of the organic molecules that make up a sample; the higher the temperature, the greater the molecular movement, and the greater the rate of degradation.

With more than three-quarters of the spots in the cryobank currently occupied, Jacquie faces some tough decisions regarding the value placed on specimens. The criteria for determining a specimen's value are generally the rarity of the species, how useful it might be to scientific research, and the quality of the sample. But a lot of this value is purely economic. Jacquie tells me that she now has to essentially run the cryobank like a business:

> So we're quite an old fashioned institution, but sadly we've been dragged into the sort of business side of things, the cost-per-sample business model type of way of running biobanks, and I'm particularly bad at it! I need to retire and let someone who knows what they're doing take over. . . . We don't say "profit," but we have to recover costs, because it's very expensive—our carbon footprint is huge, the electrical costs are huge. There's quite a problem around how we sustain this, who pays for the costs? Keeping things frozen at those ultracold temperatures is extremely expensive.

She goes on to tell me how government funding is drying up, grants from research institutions are often short-term, her staff are retiring and not being replaced—all this has resulted in the cryobank having to recuperate its costs. Only one of the three liquid nitrogen tanks is in use because of the expense of keeping them running; I am expecting an astronomical figure, but Jacquie quotes me around £5,000 a year. That one of the most visited museums in the world is struggling to find such small amounts in order to safeguard thousands of rare specimens is shocking. Jacquie admits her frustration with trying to do such important work without the help she needs. She shows me a robot that can reorganize and barcode ninety-six test tubes at a time—a potentially useful device, but because it needs to be manually programmed, it remains chronically underused. There is also an ambient storage room for dried specimens; Jacquie believes this is the future of biobanking—preservation without the reliance on electrical freezing systems that might break down—but most of the room's specially designed holding cabinets lie empty.

The thread of life does not exist in a vacuum; it is a product of a carefully honed balance of vested interests, commodification, and

Liquid nitrogen tanks are the best freezers for the cryopreservation of DNA and cellular material, but they are more expensive to operate than traditional freezers. This is the only liquid nitrogen tank currently in use at London's Natural History Museum.

surplus value. The new technologies of cryopreservation and its potential for de-extinction are funneled through this system, underscoring a process that converts raw genetic material into something more manageable—and bankable. Essentially, as Jacquie says, the DNA has to produce some form of value. What becomes bankable is compounded by the threat of extinction in the Anthropocene, now at a much higher rate than what is considered normal. This is a brand-new type of conservation work, in which the difficult realization is that perhaps there might not be time, at least right now, to save some of the creatures that are most critically endangered in the sixth great extinction. Instead of focusing on living and viable populations, the race now becomes one of gathering salvageable genetic material, either for use in captive breeding or artificial insemination programs or as safeguards for future de-extinction schemes. These cells, genes, pieces of flesh, and plant life then take on an altered state as frozen life, suspended and cataloged until it is needed at a later date. Here, the cryobanks become archives of both past and present, encapsulating the Anthropocene anxieties of a rapidly warming world and a meteoric rise in species extinction.[28]

The developments in genetic and cell biology during the past century have culminated in the cryobank; it is not too much of a stretch to understand the trajectory of the mammoth from prized museum skeleton to plastic vials of tissue. This is a process of molecularization that renders life more controllable in the cryobank and confers a great deal of power on the people who have access to the genetic material.[29] In particular, the push for the mammoth's de-extinction places emphasis on the genome and on the ability to control and manipulate laboratory tissue cultures outside the individual living body (which, of course, no longer exists). The genetic features that "make" a mammoth—shaggy fur, small ears, cold-adapted blood—take on a greater importance than the sum of their parts; it is only by replicating these species indicators that a mammoth might be said to be resurrected through hybridization.[30] And similarly, in the case of Hwang's cloning efforts, the intact cell plucked from the decaying body of a dead mammoth becomes the locus of life—the possibility of coaxing cell reproduction in a laboratory that might one day restore an entire species from extinction.[31]

The Cryopolitics of Future Life

Most mammoths are discovered by tusk hunters or by nomadic/seminomadic Indigenous reindeer herders who traverse isolated patches of tundra and thawing permafrost. Scientists rely on tips from these groups, often striking uneasy bargains in which they agree to look the other way regarding some of the more illegal practices of tusk hunting, for example, in exchange for access to the mammoth body (minus tusks, of course). Sometimes tusk hunters are directly employed by scientific institutions, as in Sasha's case, while some become makeshift guides for visiting researchers, such as the delegation of international scientists who were filmed for the UK Channel 4 documentary *Ice Age: Return of the Mammoth.*[32] One scene in this film shows the scientists wracked with anxiety as they watch the tusk hunters' high-powered water cannons blasting through the permafrost, terrified the jets will damage the fragile tissue of preserved flesh. The mammoth body in this arena is an unstable one, still attached to organs and bones, too heavy and unwieldy to move. She occupies a liminal material state, not quite frozen solid but not yet a festering mass of rotten flesh. With the mammoth not yet molecular *or* frozen, this is a period of tension, with vested interests clashing, the time for decision making truncated by rapidly rising temperatures. The mammoth body must be transferred from this difficult space as quickly as possible into the controlled environment of the freezer. Only then can she become molecular.

I am keen to go with Sasha on a tusk hunting trip, and there is a point at which it looks like this might be possible, but eventually he deems the winter too advanced. Mammoth bodies have been dug and blasted out of frozen permafrost (the Jarkov mammoth springs to mind), but it is much easier to find a body when the softening and retreat of thawed earth have done much of the work for you. Of course, this method comes with the risk that the creature is already on its way to decomposition—not an issue for tusk hunters, but a huge variable for the scientists who seek to gather viable genetic material. Here, the melting body is in a state of emergency, almost as if its frozen status delineates its healthy life, and any sign of the deadly forces of thaw necessitates a mad ambulance dash to the operating table, where it can be saved through the curative properties of freezing. These mammoths are dead organisms of course, but locating life at the molecular level blurs the traditionally distinct categories of life and death. As Adam Searle states: "The liminality of de/extinction *in the present* is taking-place and taking-shape in numerous contexts as a genomic library of nonhuman life; while technologies and imaginaries emerge so too are conceptualizations of immortality."[33] Whether this be a vial of blood or a hunk of flesh—or, indeed, the whole mammoth body itself—as long as the forces of decomposition on cellular life can be halted, the resurrection of the mammoth becomes less a necromantic fantasy and more a scientific possibility.

But the anxieties of melting are never far away. Jacquie recounts a story of collecting snail samples in Vietnam and freezing them in the field with dry ice. The research team's flight back to London was then rerouted because of an incoming typhoon, and overzealous border guards delayed the group at customs—all the while their precious cargo getting warmer and warmer. She recalls the tense flight back to London:

> We drank so much gin on the way home, and then landed in Heathrow eventually, delayed! My husband picked us up from Heathrow and I said: "Put your foot down! Straight to the museum!" Pulled up outside the museum here and then they wouldn't let us in! Like "Who are you?"! I'd asked them to warn security that we were coming late but the message hadn't got through. So they wouldn't let me in, and there was another delay while they sorted all that! But we got the boxes back in here and it was still frozen when we opened it. We were leaping around like little maniacal Rumpelstiltskins in joy!

The thread of life was allowed to continue in this case, but it was touch and go. The stretchiness of time in the cryobank became suddenly less elastic, and with the fear inspired by a rapidly emptying hourglass, the threat of melting loomed ever closer. Again, the homogeneity of time suspended through freezing is punctured by discontinuity and unpredictability—it is never completely clear how much time one might have and how much damage melting might do. With the mammoth, things are a little more complex. Hwang's failure to find any viable cells for cloning has nothing to do with the success of his smuggled cooler operations, but rather with the unpredictability of melt, which became apparent only at the molecular level. The horse we met at the beginning of this chapter, lying on the gurney so lifelike and so fresh that Semyon might have been tempted to sample horse tartare, disguised the decay of its cellular tissue well; the unpredictability of permafrost had done its work once again, the damage discovered only once Hwang ran his experiments back in Seoul.

The molecular state is facilitated by the scientists at the Mammoth Museum. Here, the rapidly thawing mammoth is transferred to the walk-in freezer, a space certainly not monitored and certified like the NHM cryobank, but one that delays the processes of melting nonetheless. The creature will be assessed and dated, possibly autopsied, her state of preservation and potential for viable DNA becoming the catalyst for the booking of flights and the convergence of other scientists from faraway laboratories. Hwang has first dibs, of course, although the documentary footage of Buttercup's autopsy depicts scientists from both the Sooam and Church laboratories—including George himself—hunched round the table, sharing pieces of flesh and vials of blood.[34] Given that the de-extinction method using gene-splicing seems better poised for success than Hwang's cloning technique, let us follow the now pocket-sized mammoth bits across the ocean to Church's lab at Harvard University, where they become further refined by the technology of CRISPR.[35] The molecular mammoth here is fully reduced to her vital components, the potentiality of her life found within the smallest of scales and the whims of scientist and machine.[36] The control and suspension of life in this arena align with the ideals of the Church lab and its partners the Long Now Foundation and Colossal—that de-extinction is a practice that we, as humans, must undertake within our remit as the eponymous masters of the Anthropocene. This almost Promethean endeavor, in which the fire is now ice, bestows cryopreserved and controllable life on a planet experiencing its sixth great extinction.

An interesting thing happens to time here, too. As Abou Farman observes: "Suspension is characterized by the translation of life into time, and its possibility produces new senses of time in life, beyond death."[37] Preserving specimens in the cryobank—particularly specimens that are already centuries old—suggests a temporal lengthening, a sort of plasticity that can be stretched out indefinitely while staying very much the same. It suits proponents of a "good Anthropocene" very well that time might be controlled, and potentially even reversed, through freezing technologies. That global temperatures are rising and the Arctic is melting is much easier to come to terms with if human-controlled freezer systems offer a safeguard. It is also reassuring that one might, to borrow Joanna Radin's term, put "life on ice"; perhaps we don't have the solutions to save X *now,* but we might in the future.[38] This is the model most cryobanks work with. This is the model that de-extinction abides by.

This process of molecularization and subsequent de-molecularization of the mammoth through various vested interests is indicative of this grasping for greater control as we hurtle toward a future that portends a lack of it.[39] Like the turning point Landecker identifies in scientists' ability to further manipulate life through laboratory techniques such as tissue culturing and freezing, proposing the de-extinction of the mammoth is an extension of this lust for control.[40] The ability to manipulate and resurrect genes and long-gone creatures suggests that we might "swap" (just as CRISPR does) the looming terrible futures of uncertainty for one of benevolent management and stewardship. While the vision of the Pleistocene Park might be to allow the techno-mammoth to wander around and cause havoc in relative freedom, the creature's attachment to the interlocking forces of cryopolitical molecularization and genetic manipulation means that notions of "bringing back" are imbued with the meddling hand of the Anthropos. The mammoth's task is no longer to *be* a mammoth, whatever that might be; now she is burdened with an altogether weightier labor—to right the wrongs of humans who refuse to relinquish control of, and over, life.

Immortality and Resurrection in the Secular Age

De-extinction is not merely the "resurrection" of individual animals; it represents a fundamental disruption to the ordering of life, death, and extinction as a linear path. It offers a reversal of death; it offers, in essence, immortality. There is a long and perhaps surprising (to an outsider, at least) history in Russia around resurrection and immortality, even within the natural sciences.[41] One particular scientist of note to this study must surely be the "father of permafrost" himself, Mikhail Sumgin. Sumgin imagined a museum he called the Subterranean Museum of Eternity, for which he would gather every organic specimen on Earth to be preserved in perpetuity in a giant permafrost repository, where future scientists could study the samples without fear of their ever spoiling or changing state. Sumgin's work aligned well with Bolshevik idealism and the Soviet push to master the North, and this contributed significantly to the Soviets' official adoption of his spatial and temporal understanding of *vechnaya merzlota*.[42] Despite resistance from fellow scientists, the designation of "eternal" could mean, in Sumgin's own words, "all frozen earth of an uninterruptedly enduring nature, from two years up to the time of the mammoths."[43] As the Soviet Union built eastward and northward, the fact that eternity could easily mean a mere two years was integral to the project of mastering time and space.

But Sumgin's museum was much more concerned with permafrost's *potential* for eternity. It began construction in the 1960s in the little Arctic settlement of Igarka but was never completed. Today, Igarka is yet another dying Siberian town in the vein of Chersky, with the half-finished museum the only entry on its Tripadvisor page. Possibly the earliest archival cryobank in existence, it contains a paltry percentage of the bounty Sumgin had imagined, and what might happen to its meager contents as the permafrost around it continues to thaw remains unknown. Despite Sumgin's intimate knowledge of permafrost, he did not entertain the notion that this "natural" cryobank could fail:

> We are talking about an institution that must work for thousands of years without the slightest interruption. . . . We cannot guarantee that the refrigerator on the ground will work without interruption. . . . By contrast, a refrigerator built in permafrost soil will work in the sense of preserving the contents unconditionally without interruption for thousands of years.[44]

His words stand in direct contrast to the current rush to fill electrically powered freezers and cryotanks as the planet continues to warm.[45] In the Anthropocene, it is the failure of the permafrost to continue its permanence that necessitates the drive toward more controllable safeguards; here, Sumgin's notions of immortality and eternity are flipped on their heads, as the trust he placed in the longevity of permafrost erodes day by day. It is now the freezer that houses the possibility of eternity—the idea that as long as humans can keep the machines running, the thread of life might continue indefinitely. But the Soviet fantasy of immortal permafrost wrenched painfully into the future lays bare the very *im*permanence and fragility of preservation by freezing of *any* kind; indeed, in the case of the NHM's cryobank, use of the liquid nitrogen tanks rests on the ability of one scientist—Jacquie—to balance the budget.

Taken this way, it becomes apparent that the motivations behind Sumgin's museum are little different from those behind the modern cryobank, in that both institutions cement the mastery of human science, albeit through different material contexts. Sumgin fantasized that his museum would become invaluable to science in the future, wondering whether it would be possible to "make sure that scientists who will live tens of thousands of years after us could study the animal world that inhabits the earth at the present time with the help of permafrost."[46] That the immortality of the bodies locked in permafrost took on any significance only once humans deigned to study those bodies chimes with the motivations of cryobanks around the world today. It is not about the eternal life of creatures like the mammoth at all, but about the immortality of *humans* as a species and the endless pursuit of mastery over science, knowledge, and life itself. In a period in which the looming threat of apocalypse—essentially, the death of human time—implies a potential human extinction event, the fact that freezing technologies can extend life almost indefinitely means that the notion of finitude takes on a sort of malleability. It is not just life that becomes located within the frozen cell, but the very survival of humans on a planet getting hotter and hotter.

The de-extinction of the mammoth is part of this survival strategy. As a keystone species in the Pleistocene Park's restoration program of permafrost thaw mitigation, the mammoth plays a role that has less to do with the resurrection of her *own* species and more to do with the perpetuation of the human—or, rather, the perpetuation of an extractive, capitalist-dominated way of life that cleaves to geoengineering strategies for fixing the problems of the Anthropocene rather than addressing their causes. While the park styles itself as a holistic and natural endeavor, the Zimovs also do not conceal the fact that they do this work for the survival and benefit of humanity. Meanwhile, the proponents of mammoth de-extinction—particularly those funded by tech bros in Silicon Valley—fully ascribe to a good Anthropocene in which a techno-utopia is made possible through for-profit science schemes. The return of the mammoth would crown this achievement as an almost literal embodiment of resurrection and the final piece of the puzzle in returning to life a lost ecosystem that will, according to the park's mission statement, save the world.

This drive for immortality and resurrection is nothing new, of course. It is difficult to ignore the religious undertones that inflect all these endeavors, right down to their names—a couple that spring to mind are the Lazarus Project, which sequenced the genome of the southern gastric brooding frog, and the Frozen Ark cryobank.[47] But while many of these projects are criticized for "playing God," religion has very little to do with it. The eschatological time—the destiny of humanity—of the Anthropocene is not defined by a passing into the afterlife, but instead is contextualized by the possibility that humanity might be coming to an end *without* redemption. This is an apocalypse without a higher power, in which the horror of human time appearing to run out thanks to the destructive practices that produced the Anthropocene generates schemes to stop, prolong, and even reverse time. What Farman terms "secular eschatology" occurs as a result of humans attempting to extend their survival on a planet that is becoming increasingly hostile to them.[48] He points out that the anxieties of death in the secular age come not from believing that there is no afterlife, but rather from the fact that everything else continues without you. Apply this to the human species and the threat of extinction—there is nothing to indicate that the planet will not just continue on without us, and therein lies the real terror. The construction of the permafrost (and the Arctic in general) as a "ticking time bomb" is indicative of the notion that our time is running out, and that something explosive is waiting at the end.

The glossy website of the de-extinction organization Colossal declares the mammoth to be "Earth's old friend and new hero."[49] Disregarding the quite patently false claim that the mammoth was ever a "friend" to humans, the future mammoth is constructed through the logic of secular eschatology as a hero for humanity. The task of the Western-science-based, largely male-dominated proponents of a good Anthropocene seems to be to use de-extinction and resulting geoengineering practices as a way to extend and ensure mastery over both life and time, thus suppressing the unsettling material forces of thaw that might threaten the relentless march of "progress."[50] Shifting the control over life from the permafrost to the freezer doubles down on this control—despite the potential for freezers to fail, the instability and unpredictability of a warming planet necessitate a greater urgency in the quest for both human survival and dominance. Within the cryobank, the finitude of life is materially reorganized through the forces of nonlife to make it theoretically infinite—the suppression of an unruly Earth. This is not, in fact, the erasure of God, but merely the erasure of the mystique of an invisible yet omnipotent presence, as the Anthropos steps into the vacant role of all-encompassing mastery. Blood, once again, makes a return through the ritualistic practice of collection and storage; the mammoth's resurrection is made possible through the banking of her body and blood. Time is desiccated and put on ice, ready to be deployed through the redemptive possibility of the mammoth. Her molecular body is made flesh, and, again, I am reminded of this quotation: "We are as gods, and we have to get good at it."[51]

Giving Up on Life

De-extinction science—like all new scientific paradigms—has not simply emerged from nowhere, nor has it expanded without a specific goal in mind. De-extinction and cryobanking are responses to a conservation emergency in which it is impossible to prevent species from going extinct at the current rate. But just as with Jacquie's choices at the NHM cryobank, who gets to decide which creatures are valued as de-extinction candidates—and, indeed, who might have access to the sort of funding necessary to continue such a hugely expensive project—is subject to the whims and motivations of what Marisol de la Cadena and Mario Blaser term "the world of the powerful."[52] Ben Novak, one of the stars of Stewart Brand's Revive and Restore de-extinction project, became obsessed with the passenger pigeon at a young age after seeing taxidermy specimens in a museum—so much so that he made it his life's work to resurrect the bird.[53] George Church is a highly controversial figure who finds no ethical quandary in human gene editing, in bad company with Hwang Woo-Suk's earlier work.[54] This is profoundly masculine science, in which the structural power of maleness works with the very "macho" nature of experimenting with pioneering and often quite fringe scientific practices.[55] In a disparaging write-up of the Pleistocene Park project, Vincent Bruyere calls the park the "site of an open-ended experiment in storytelling set in a climate of uncertainty, where men have vouched to give birth to mammoths, which will in turn ensure that men have a future to remember."[56] This faintly ridiculous visual somewhat eclipses the grain of truth found in his statement: that birth, rebirth, and reproduction are not merely biological processes but also can be wielded as tools of mastery and control. The Pleistocene Park's logic is predicated on the rewilded animals' reproductive ability to save the world for the *human* children of tomorrow. The idea of reproductive futurism is mobilized often in climate change discourse: that we must think of the children before we trash the planet. Not only does this disregard other forms of becoming and kinship, but it also reinforces the heteropatriarchal narrative of apocalyptic manhood.[57] It is a future of men to be saved.

This also extends to de-extinction. While the unpredictable earthly forces of thaw have generated this cryopolitical turn, de-extinction works with the premise that life cannot reside in the controllable spaces of the laboratory forever. At some point, for success to be reached, the de-extincted creature must become de-molecularized and "go outside," as it were. Returning to the idea of blood, there is a strange tension found in the potentiality of forty-thousand-year-old blood to produce a new mammoth with no familial kin or blood ties; blood is, as Sarah Franklin states, "thicker than genes," and this statement surely queries the familial bonds that might form between de-extincted mammoths.[58] Resurrecting an extinct creature requires a birthing, or perhaps a rebirthing, of the first new baby of its species, bringing with it new and awkward forms of kinship.[59] In the absence of mammoth parents, what might constitute a new mammoth's lineage? Can it even be called a mammoth? And what of the task of reproducing more mammoths, if there is initially only one in existence? If the masculinity of de-extinction might encompass the role of the creature's father, spawned through molecularization in the human male spaces of the laboratory, then there is a solidification of the constructed boundary between nature and culture once the baby mammoth enters the feminized space of the outside—and now terrifyingly unpredictable—environment. With any luck, the tottering infant's rage will be placated and her "natural" propensity for motherhood will be stimulated, with the Earth returning to her role as nurturing life-giver; the mastery of the powerful is now complete.

How to challenge this? How, indeed, to put a stop to this continuous birthing and rebirthing of the destructive practices and unequal structures that underpin the promotion of de-extinction science and cryobanking? This is not to say that many of the scientists and conservationists involved are anything less than dedicated people trying to make a better world, but what they often fail to understand is that attempts to suspend, prolong, and perpetuate the particular way of life (and life itself) that brought about the Anthropocene in the first place will never fix its brutalizing effects. In a wonderful polemic on queer futurity, Lee Edelman points to the sanitized image that places children—and by extension the heteronormative white family—as our future as an example of what he calls "coercive universalization," which categorizes difference as a dangerous enemy to be destroyed.[60] It is an easy notion to cling to—who could possibly be against a better world for the innocents?—and one that saturates de-extinction, with its narrative of lively resurrection and (re)continuity of species lines. Despite advances in genetic science through CRISPR and the molecularization of life, for de-extinction to be a complete success, eventually the propagation of cells must once again take bodily form, and the resulting bodies must engage in sexual reproduction. This sort of reproductive futurism is intricately bound up with the temporal linearity of human continuation, in which a feminized Earth does what she is told and provides a safe world in which our children can grow up.[61] These future children, of course, are whitewashed by racism and banality, while Black, brown, and Indigenous children experience current and ongoing threats to their survival.

There must be a way out of this destructive cycle, a way that allows for the differences that characterize the multiple planetary scales, temporalities, and ontologies that make up permafrost worlds and beyond. I would argue that for the cycle to be broken, the continuous thread of life must be destroyed. While this may seem like a radical concept, it can be unpacked in such a way that discontinuity is allowed to enter any fraught discussion around survival in the Anthropocene. I am not advocating for the extinction of the human species, nor am I promoting discontinuity as a panacea for the violence meted out by the world of the powerful. Rather, discontinuity is open to surprise; it disrupts the valorization of the nuclear family and makes (discontinuous) spaces for a queer heterogeneity.[62] The discontinuity of permafrost so feared by the powerful can be a blueprint both for new futures in which attempts at mastery are resisted and for a new dynamic material engagement in which we become more mindful of earthly agency. This involves refusing any fantasies of immortality or redemption through resurrection. Instead, the task is to practice discontinuity as well as recognize it, through disrupting and protesting the hegemonic linearity of the powerful. Through this emerges a commitment to a sort of "giving up" on life—not in a way that accelerates the demise of humans on the planet, but rather in a way that works to discontinue the continual birthing of the kinds of power structures that underpin the Anthropocene.

The Lives of the Future

On my final day at the Pleistocene Park, I wander down to the banks of the Kolyma to check out the tank. Nobody can really remember—or they won't tell me—how Sergey managed to get ahold of this hulking Soviet lump of metal, or how he got the thing up to the Arctic, but manage he did, and now it sits gently rusting in the river mud. It has not been working for a while, but when it was, the Zimovs would use it to mimic the destructive capabilities of a mammoth—trampling and compacting permafrost soil, knocking over trees and shrubs, playing general havoc with the ecosystem. One of Nikita's favorite ways of catching journalists' attention is to proudly renounce his status as an environmentalist by explaining how much he loves destroying the landscape. "According to the vision of absolute majority of people in the world, I just did a horrible thing," he declares. "I was driving around the forest intentionally killing trees!"[63]

While the tank's existence points to the very masculine (and stereotypically Russian) frontierist science undertaken at the park, it also serves as a stark reminder that stand-ins are necessary to restore the Pleistocene ecosystem in a world with no mammoths. The park is not expecting a mammoth anytime soon, and must compensate for such a creature's destructive capabilities with technology

In the absence of any "real" mammoths, the Pleistocene Park used this Soviet-era tank to mimic their destructive nature by knocking over trees and making space for grassland to colonize.

(albeit clunky and rusting in this case). The "life" of the mammoth currently exists only in vials of blood and clusters of cells at subzero temperatures, as yet unable to make the leap into fully realized beast with a mission, while the planet continues to get hotter. To carry on fighting the ever-encroaching permafrost thaw, and the tundra vegetation that exacerbates it, the Zimovs need to fake it; the "natural" geoengineering strategy they claim to promote is a long way off. Sergey struts around the tundra in a T-shirt sporting a mammoth's likeness. "I need maybe fifty thousand mammoths," he declares. "But no one [will] give me so many!"[64] The imagined mammoth takes on the form of planetary savior, of a behemoth emerging from the mists of time to guide us back to safety, a ghost on the tundra of things that could be. This is a future that harnesses the simple power of the gene imaginary: that life—and through life, redemption—emerges from a simple cluster of cells. If humans can just manipulate these cells and genes to their advantage, the Anthropocene and its apocalyptic futures might no longer seem so threatening.

The relocation of life from the organism to the molecular level through laboratory freezing techniques is an attempt at shoring up another layer of control on an increasingly unpredictable Earth. To cryobanks like the NHM's and the Frozen Ark, collecting genetic material to store in freezers is a safeguard against the spiraling rate of extinction; their organic repositories keep samples suspended in stasis until a time when they might be needed for more practical measures.[65] To de-extinction scientists such as George Church and Hwang Woo-Suk, the ability to replicate and preserve life through cell cultures and genomes represents an opportunity not only to renegotiate the properties of life but also to create life. The cryobank becomes a tool through which to extend mastery and control over life, the freezer acting in direct opposition to the unpredictable forces of melting that rule beyond the bounded and continuous space of the laboratory. The thread of life that exists through cell culture across generational, species, and organism lines is made possible by the preservational properties of freezing, through which life is suspended within an ambiguous cryopolitical space between being alive and being allowed to die. Blood, in both its material and metaphoric forms, becomes imbued with lively potential and disrupts temporalities, perpetuating the linearity of unbroken bloodlines while also getting caught in the temporal loop of resurrection. In this sense, it becomes possible to do away with the inherent finitude of life and achieve a sort of immortality—a secular eschatology that abandons the unpredictable notion of God, recasting humans as ultimate masters of their own destiny as a way to counter the growing fear of apocalypse and extinction.[66]

The mammoth is wrested into this arena from her millennia-long home in the permafrost. Her potential resurrection emerges from the uncovering of her dead body through the retreat of matter and the resulting anxieties of extinction. Once revered in her full form in museum exhibits and scientific specimens, she has been steadily broken down into her constituent parts until she embodies the molecular, chopped up and put into freezers, squeezed into test tubes and vials, shipped off to destinations far from her original resting place. In becoming molecular, she is imbued with lively possibility, her cells and genes entering technological spaces that stretch and enlarge, both materially and temporally. The life encased in her cells is temporally translated, encompassing the vast reaches of her history and possible future while simultaneously suspending time altogether. But beyond the boundaries of the cryobank and the cell wall, time has other ideas. The unpredictable forces of melting, through which nonlife infiltrates life, produce ruptures in the continuity of life—discontinuities that disrupt the linearity and speed of human-constructed time. The life of the molecular mammoth might be safe (for now) in the cryobank, but the conditions of her resurrection involve a task: to restore the thawing permafrost that was once her tomb and resist the apocalyptic wave of time crashing back on us.[67] The very impermanence of permafrost is destructive of both past time and future time in which the human species gets to continue on and on, generating a fear that time is running out, and resulting in the desperate push to freeze and control time through cryobanks.

The Zimovs drive their tank over the permafrost for this reason—because they are running out of time. Over in the NHM, Jacquie bemoans how little time she has to configure the cryobank as she wants it, and worries about the longevity of her work: "Museum biobanks are really, really long time frames, hundreds of years, so we really are thinking 'forever' . . . that's the thing about being at the museum, you're only there for a short period of time." Brand and Church's de-extinction endeavors form part of the remit of the Long Now Foundation, dedicated to bringing about long-term change in a world of immediacy and short-term thinking. Hwang races against the clock as he attempts to restore his respectability and cement a legacy of scientific greatness. These intersecting and often differing motivations converge in one very obvious junction: that the paradox of the Anthropocene attempts to maintain dominion over both life and time in order to prevent the destruction of humanity. The mammoth's de-extinction might be the crowning achievement of this objective, but as Ross MacPhee, curator of the American Museum of Natural History in New York, forcefully stated when he participated in a 2019 debate panel about de-extinction, all it amounts to is "hubris, hubris, hubris!"[68]

It surely does not have to be this way. By abandoning the valorization of the replicating subject, and thus of life, we can find other ways that refuse the perpetual hunt for mastery. The fantasy of control that the Anthropocene perpetuates is ultimately just that: a fantasy. Or, as Daniel Cunha states, the Anthropocene "is controlled neither by humanity (anthropo) nor by a part of humanity (the ruling class) . . . it is a situation increasingly *out* of control."[69] De-extinction is touted as a way to clean up the messes left behind by humans; sometimes, it is even constructed as humankind's moral responsibility to the species concerned—we killed you off in the first place, now we have the means to bring you back. But the promise of a "good Anthropocene" is a poisoned chalice if it is not accompanied by the dismantling of the destructive practices that created the sixth great extinction. The mammoth's resurrection and rebirth sees her become part of a narrative that requires labor on behalf of this controlling Anthropocene subject: for her to produce a legacy *not* for her own species, but for the continuation of the powerful. This is a profoundly natalist vision, in which the thread of life through birth is made possible by the manipulation of frozen genes and reworked blood ties. Birth, and rebirth, is inextricably bound to the Anthropocene narration of the future; hand-wringing and cries of "Won't somebody *please* think of the children?" are funneled through the expectation of mammoth labor to produce an offspring of hope. Maintaining human reproductive futurism *through* new mammoth lineages denies the mammoth her right to rest in discontinuity.

In the world of the powerful, de-extinction becomes mobilized as a tool of mastery, immortality proffered as a magic bullet that destroys the idea of death and loss. I propose that instead we stay in the discontinuous spaces that extinction produces—those absences and unravelings that are reminders of the assigned hierarchies of life. Permafrost can help. Its material and temporal unpredictability is a refusal to conform to a continuous renewal of Anthropocene dominance. Instead, it produces cracks that reveal the fate of the mammoth through her history as an embodied, storied beast whose life was more than what her genes might offer. Buttercup exists both forty thousand years ago and now, as a body plucked from permafrost and held in the Mammoth Museum's freezer. Her body is a point of discontinuity that marks the places where humans have used her for samples: the holes in her side that gushed blood, the scalpel slashes from her autopsy, the pieces missing, taken from her and shipped all over the world. I suggest that, rather than attempting to reproduce her form in the cryobank and maintain the thread of life through her frozen genes, the proponents of a "good Anthropocene" should perhaps take a moment to sit with what it means to destroy extinction, what it means to manipulate time, and what it means to attempt to master life and death. The answers, like permafrost, may be surprising.

Conclusion

In 1915, the Russian geologist Vladimir Obruchev wrote *Plutonia,* a science fiction novel that follows an expedition of eccentrics to the Russian Arctic, where they come across a strange depression in the tundra.[1] As the group stumbles through the thick fog that shrouds the inhospitable landscape, they feel themselves descending gradually into the belly of the Earth. After a while, they glimpse what look like grassy hillocks moving slowly in the distance, a sight that they initially attribute to the strange land in which they find themselves. A glance through binoculars, however, reveals the hillocks to be elephant-like creatures—in a word, mammoths. The immediate impulse of the group, as with most colonizers, is to shoot first and ask questions later. They kill an animal that does indeed turn out to be a mammoth, and this incident sets the scene for the rest of the book, as the group encounters—and kills—an array of prehistoric fauna during their journey deep into the bowels of geological history. At the final level of this hollow Earth, they find a tribe of "savages"—prehistoric humans—who capture two of the explorers and hold them hostage until the others use their guns to terrorize the tribe and rescue their friends.

The book follows a rather unoriginal format of dashing explorers using the logic of science to explain their surroundings—a trope frequently applied to the Arctic, notwithstanding any fantasies about hollow Earths—while projecting their ideas about "civilization" onto the native population. The explorers' journey mirrors the trajectory of the Russian movement east and northward and the colonization of Siberia and the Arctic, where Cossack guns came up against Indigenous spears and bows.[2] *Plutonia* does end on a glum note for its protagonists, possibly a warning from Obruchev for all would-be scientific explorers. After the sea ice melts, the group is able to sail away from the Arctic, only to be captured by an Austrian warship. They learn that Russia has been at war with Germany for an entire year, and they are now prisoners of war. They are able to convince the Austrians to let them go, but they lose all of the specimens they had collected as proof of an inhabited hollow Earth. Their scientific endeavors come to naught, and nobody believes their story.

The story of *Plutonia* contains many layers, many intersecting narratives and histories, much like permafrost itself. Despite its derivative structure and subject matter, the novel offers telling insights into how Russian scientists viewed the Arctic on the eve of a revolution: as a land full of strange and wonderful beasts ripe for the shooting, where people who have not yet had "civilization" foisted on them eke out a meager existence in the frigid soil, an inhospitable land of swirling winds and brutal snows almost asking to be conquered by intrepid explorer types. This understanding of the Arctic was carried forward as the Soviets pushed on with their plan to "master the North," with permafrost subsumed into this narrative as something to be tamed, controlled, conquered. The budding discipline of permafrost science was built on the idea that permafrost is stable, permafrost is permanent; entire cities were built and huge mines were dug, buttressed by a seemingly endless supply of forced labor. When permafrost began thawing beyond what was considered normal, this carefully constructed—but fragile—assumption that the eternal frost is indeed eternal started showing cracks.

Cracks can take many forms. A crack in a windshield can go unnoticed, sometimes for years, before the glass shatters—the crack causing the gradual weakening over time of something assumed to be solid, until it isn't. A crack can be large enough to fall into, perhaps a crevasse or a cave—such cracks thwart the notion of stable ground, they puncture the skin of the Earth. Cracks happen in ice; if you try to drive on the Lena River ice road too soon, as one hapless driver did during my stay in Yakutsk, you may hear the ominous cracking of not-fully-formed ice beneath your wheels (the driver survived; his car did not). Permafrost cracks, just as it slumps and subsides and breaks and turns to jelly and mush, retreats and gives way, making space for other things. Cracks in the earth are spaces where air or water can rush in, evidence of a spatial transaction of materialities similar to the way a pair of lungs work. Permafrost has always, to use the lexicon of Nikolay Sleptsov-Sylyk, breathed, the undulating freeze and thaw of the active layer part of its seasonal rhythms; now, however, permafrost might be said to be gasping.[3] Permafrost expert Merritt Turetsky has also used the term "breathing" to describe permafrost's dynamism in the Anthropocene, from a perspective based in Western science.[4] The land heaves and collapses under the strain of thaw, belching carbon dioxide and methane out into an atmosphere already saturated with them.

The notion of the crack demonstrates the discontinuity found in permafrost and the existence of a discontinuous Earth. The planet has always been discontinuous, of course, but its current Anthropocene imaginary as a "whole Earth" populated by homogeneous species groups, the most powerful of which—*Homo sapiens*—is responsible for the geological epoch many argue we currently find ourselves in, means it is more important than ever to pay attention to its heterogeneity. What the flattening of the Earth does is offer the illusion of human control, the shiny surface of a globe an exquisite pretense that the planet stays the same, forever predictable. Permafrost becoming more discontinuous with the advancement of anthropogenic climate change and other destructive practices produces a tension between human and planet—not a dichotomy, but an antagonism loosely held together by a network of multiple agencies and power differentials. Found within this tension is the anxiety of extinction. The notion of apocalypse is keenly felt the more the Anthropocene and the "whole Earth" idea bed in; rather than addressing the ongoing and current environmental destruction that threatens both livelihoods and life itself, we develop a fixation toward an apocalyptic future event that will destroy the increasingly tenuous surety of anthropogenic dominance.[5]

Permafrost unsettles the fantasy of smoothness and linearity associated with living on the surface.[6] Discontinuous permafrost is a perforation, slumping and slipping and giving way into cracks of all sizes—reminders that the Earth is actually, if you look closely enough, a holey place. This is something that humans have not particularly liked to think about too much, and so the underground is often ignored or rendered as a horrific other in relation to the safe space of surface life. The stories of *Plutonia* and other hollow-Earth novels act almost as grotesque displays of the planet's insides—something we are not actually meant to see. We feel more and more uneasy the farther into the depths of the hole the protagonists go, and while what they find down there is quite obviously science fiction, there remains a kernel of discomfort. The Earth *does* enclose the remains of lost worlds, lost ecosystems, the bones of extinct creatures. The "savages" encountered in the final level of the hollow in *Plutonia* are early humans who have similarly gone extinct—a warning to the explorers that no matter how much they shoot and kill, or gather scientific evidence, the specter of human extinction will never quite go away.

The course of this book has come a full circle, albeit a discontinuous one. The prospect of human extinction has taken many forms across the reaches of human history, woven into tales, folklore, and scientific discovery, and more recently becoming a big-budget draw in literature and film.[7] I began by interrogating how permafrost thaw adds to the global apocalyptic narrative of runaway climate change, scaled up, with nuance and difference discarded in favor of a vague notion of "permafrost" on a map. I ended by addressing the prospect of de-extinction, a seemingly easy fix—if an ethically dubious one—for the careening rates of species extinction across the planet and a possible way of geoengineering and mitigating permafrost thaw. While the mammoth is placed front and center in the de-extinction drive, what is revealed upon a little deeper digging is that resurrection biology is *still* about maintaining control over life and death, and reversing the threat to human survival. The doomsday narrative pushed forward by the Pleistocene Park, regardless of the work done on the ground, emphasizes the park's purpose as a planetary redemption strategy for humans. Pushing through the cracks, however, are other ways of imagining what extinction is and how it curates the messy business of living on a warming planet—particularly in the Arctic.

Think of a crack, a hole, a hollow, and what most likely springs to mind is an absence of something. This is the normative understanding of holes, but what paying attention to a progressively discontinuous permafrost does is make space not only for the absences and losses of what was once there but also for the things that might emerge from the cracks. What these things are may be surprising—ancient viruses, preserved worms, deadly anthrax, mischievous spirits or other demons. The cracks might also reveal things known to be there that have been covered until now—mammoth bodies, prehistoric ecosystems, the bones of Pleistocene animals, carbon released as greenhouse gases. Or what emerges might be completely novel altogether—new ways of living on and with the tundra landscape, alternative economies based on tusk hunting or de-extinction, a rewilding experiment on the lands of a former gulag, new patterns of breathing. These responses are often bound up within a Western capitalist system predicated on power differentials and destruction, but they are also embodied, local engagements with permafrost that require a sinking into this alien, shifting earth. What remains unchanging is that permafrost in the Anthropocene is never predictable.

Is this the end of permafrost? Is a permafrost extinction happening across the Arctic? The point of this book has been to challenge the normative definition of extinction—as the irreversible death of a species—that has come to take on so much weight in the Anthropocene. It is unlikely permafrost will ever fully disappear—at least, not in the timescales proffered by the IPCC's climate change models that speak of thresholds and temperature limits. Yet there can be no doubt that both permafrost and the Arctic are experiencing rapid heating and thawing on account of anthropogenic climate change. What exactly is being lost can be a difficult thing to define—there is no starting point, no permafrost *nascence* to hold loss and extinction against. There is simply no way of accurately knowing how much permafrost there is or has been throughout history, despite the best efforts of scientists to map and model with sophisticated equipment. Permafrost is a volume, a mass, much of it hidden beneath the surface—a pulsating slipperiness that renders pointless any attempts to quantify it as a whole. In terms of the global permafrost object, all that can be said, with varying degrees of accuracy, is that it is thawing.

But losing permafrost affects more than just the markings on a map. Dig deeper, and a network of entanglements with permafrost is uncovered, at micro and macro scales, ontologically or epistemologically distinct, encompassing multiple temporalities and different material registers depending on whether the permafrost is frozen, thawed, or something in between. Permafrost has always been a part of the lives of Sakha residents, but usually as a backdrop rather than a main player—a tuned-out hum that is now becoming a screech difficult to ignore. Sakha's permafrost law is still in the planning stages, frustratingly bound up in the workings of Russia's poorly funded federalized government system, as local lawmakers become increasingly alarmed at the lack of protection for the earth that constitutes 65 percent of Russia's landmass. Meanwhile, Russia's Arctic strategy, aggressively promoted by the Putin administration, extols the benefits of developing the Arctic.[8] The companies Gazprom and Novatek squat on Indigenous Nenet lands with their gigantic liquefied natural gas (LNG) plants, their dirty business practices polluting the tundra.[9] In 2021, an estimated eighty thousand reindeer perished because, according to Nenet claims, they could not feed after water runoff from the nearby LNG plants created an impenetrable layer of ice over the lichen.[10] With destructive developments like these that add to the climate crisis, the Arctic is becoming a difficult place to inhabit for tundra people. Yet the Russian government eyes opportunities for resource extraction and offers funding incentives to any businesses that wish to set up shop in the country's northernmost reaches.

The Pleistocene Park, ironically, is one of only two recipients of such funding in all of Sakha. Deciding on a whim to take on the impenetrable depths of Russian bureaucracy, Nikita submitted a proposal for a goat farm that would produce a trade in milk and fur. The proposal was, in Nikita's words, "a total fake"—he never had any intention of farming goats; rather, he planned to use the money, if he received it, to transport the animals to the park.[11] To his great surprise, he was awarded $15,000 as the sole proprietor of "Pleistocene Goat Farm," and the herd became the newest residents of the park. And so the cycle continues: animals in trucks, animals on boats, animals let loose in a patch of fenced-in Arctic tundra in the hope that they might save the world. The park acts as a temporal and scalar anchor of permafrost, encompassing both a global outlook and an insular, misfortune-beset locality in which the most mundane things—an animal refusing to breed, for instance—are saturated with potential catastrophe. Such catastrophes are similarly bound up in different scalar registers of extinction and apocalypse: the extinction of the Pleistocene ecosystem beneath the feet of displaced creatures and the extinction that is the decimation of Leonid the fisherman's lake of fish, uncomfortably held in tandem by the discontinuity of permafrost thaw.

The future of the Pleistocene Park is unclear. How could it not be, for such a radical and audacious attempt at planetary geoengineering? A particularly cold winter might see many of the animals die, the Zimovs could run out of money, a global pandemic might unsteady the course somewhat, a brutal war meted out by a power-crazed despot might render Russia a pariah state.[12] Yet global-scale schemes to mitigate climate change are attractive to those who adhere to the promise of a "good Anthropocene"—that our technological might as humans will save us from whatever disaster the planet might rain down on us.[13] Stewart Brand's visit to the Pleistocene Park, just a few weeks after mine, was part of a documentary film about his life as a pioneering technophile. Brand takes a particular interest in the park as a worldwide terraforming strategy, not least because his organization the Long Now Foundation advocates for and actively researches species resurrection, particularly through the work of his friend George Church. Now, more than four years later, the project to resurrect the mammoth continues, and Colossal, set up in 2021, claims that it will achieve success within five years; in the meantime, Brand's Revive and Restore project has welcomed the cloning of a Przewalski foal, born in 2020, and a black-footed ferret, born in 2021.[14] The proponents of a good Anthropocene forge forward with their commitment to the continuity of human dominance, conferring certainty on a hyperconsuming, white supremacist hierarchy that there is no need to scale back on environmental destruction, resource extraction, and colonization—technology, in the end, will save us.

The Pleistocene Park and the de-extinction of the mammoth are not merely audacious experiments, they are also radical attempts at the reordering of life, death, and extinction in an epoch of frictions and turmoil. Both work within the premise that human survival in the face of an increasingly hostile planet is a key concern, but they do so in different ways. De-extinction doubles down on the "good Anthropocene" premise that humans, as dominant geological agents, might forcefully extend mastery over earthly processes and planetary life that is teetering on precarity. The mobilization of temperature controls through the medium of freezers and cryobanks supports the very attractive notion that life not only can continue in the face of environmental destruction, unfettered capitalism, and colonialism but also can be *returned* through genetic technoscience. In a world that is getting hotter, the lure of the freezer is undeniable; the potential to create life from a cluster of prehistoric cells dug out of the permafrost is no longer only in the realm of a *Jurassic Park* fantasy—we are now, as Stewart Brand has stated so forcefully, as gods.[15]

Yet for all the park's posturing about saving the world and being the "world's best tool" to fight global warming, the Zimovs never quite detach themselves from the material embodiment of permafrost living—they are physically unable to, held fast as they are by a spatial positioning in a warming Arctic. Their vision is unapologetically anthropocentric, but this is frequently transgressed by the liveliness found within the permafrost itself—animals escaping through the fences, the permafrost tunnel flooding, wildfires that can start anywhere, soils changing with temperature and seasonal shifts. The fantasy of control offered by cryobanks and other human-operated systems of preservation breaks down beyond the walls of these facilities—the illusion of immortality is shattered by the refusal of permafrost to stay put. While the park shows some encouraging signs of responding to the Zimovs' experiment, the permafrost *outside* the park thwarts any attempts at predicting its patterns.[16] This is the crux of the Anthropocene and the anxieties of climate apocalypse that accompany it. The hubris found in attempts to restrain and configure life and survival collapses in the face of permafrost thaw, the dynamic agency of ice reformulating ontological and epistemological ways of knowing the planet.

The boundary between extinct and not extinct is thoroughly blurred by permafrost and its becoming discontinuous in the Anthropocene. In the preceding chapters, I have presented a series of overlapping stories and scenarios, heterogeneous in their scale, temporality, ontology, and materiality, to propose an intervention into the idea that extinction applies only to species life as well as the idea that extinction is forever and the sudden end of something. These are all snapshots of permafrost life that disrupt the definitions that seek to *separate* the categories of life/death/nonlife, inside/outside, and deep time/shallow time. The thawing of permafrost marks a slipperiness that means it is difficult to grasp quite what it *is* at any given time, let alone formulate a definition that might be committed to a dictionary. Permafrost confounds with its dynamism, its refusal to be categorized, and its tendency to remain hidden underground, and in doing so produces a sense of perforation—a discontinuity—in modernist assumptions of life. The designation of discontinuity proposes that extinction is not an end point or an absence; rather, it is a generative force that might take something away but also produces new ways of being with the landscape. Taken this way, extinction is recognized as an ongoing process, but not one that is smooth or linear. The jagged, rough edges of extinction in the Anthropocene transcend its globalizing moniker and bite down into the local, ongoing processes of apocalypse, the liminality of absence and presence, or the ghostly "return" of bones and bodies. Extinction as a discontinuous process is full of holes, from which things might emerge or into which things might go.

Of course, these discontinuities should be interrogated, and the power and hierarchies and entangled motivations that render so much needless destruction in the Anthropocene must be challenged. I argue that *recognizing* the dynamic force of extinction unearths these interlinked processes as ruptures, producing echoes that rumble on long after the action has seemingly finished. The Soviet push to "master the North" and the colonizing practices undertaken by the Cossacks several centuries before are inscribed in the permafrost soils and the lives that exist with tundra; they are not merely historical occurrences that altered future events, they continue to vibrate around the town of Chersky and the city of Yakutsk, and in many other tundra places beyond. They possess their own durations that push back against the supposedly linear continuity of time, drawing attention to the slower disasters and more drawn-out catastrophes that span generations—ones that are so often confined to history or discarded in favor of the grand future apocalyptic event. Allowing space for the discontinuities produced by extinction reveals the pockets of resistance, the different forms of agency and emergence that disrupt the idea that extinction is *just,* in the words of Thom van Dooren, a "slow unraveling."[17] Permafrost cracks are openings through which to understand the uneasy tension between destruction and renaissance. Nenet lands have been colonized, first by settlers, then by gas companies, but their reindeer herds have never been thicker—the point is not to deny the destructive process of extinction, but also not to deny its generative potential and the agencies of those who hold the least power. The animals in the park struggle with the weight of a brutal new climate and the expectation that they will produce a working ecosystem from the remains of an extinct one, but they might form new attachments with each other, break free from their status as planetary saviors, create new natures.

What might these new natures look like? How might they collide with the undeniable devastation being inflicted on the planet by a small subsection of the (most powerful) human populace? How might we recognize a discontinuous Earth among the homogenizing and whitewashing processes that seek to cover it? Matthew Chrulew's short story "The Mamontogist's Tale" offers a speculative glimpse of the Pleistocene Park in a postapocalyptic setting.[18] We are not privy to exactly what happened, but it appears that the planet now supports a tribe of nomadic hunter-gatherers who make references to the "teck-sapien-ape," the remnants of which betray a lost society based on technology and extraction. The language is purposely abstract, the words often familiar yet not, invoking the breakdown of the English language and its gradual piecing back together into something else, something uncanny. Mammoths saturate the story—the Russian word for mammoth is *mamont*—from descriptions of their extinction as beasts descending into "eck-stinky holes" beneath the surface to their de-extinction through "gnomogy" and "sequins." We learn that the tribe worships the mammoth by way of song and prayer, but now mamont is dying yet again. The unnamed narrator—the titular mamontogist—relates the tale of their expedition to hunt down the "sorrow gate mother" who was last seen in a place called "Plastic Park," the rusting sign for which supports two remaining letters—*T* and *C*.

Inside Plastic Park are the remains of fences and paddocks—reminders that mamont and her kin were once enclosed to be gawked at by tourists—and an overgrown tundra landscape. The mamontogist eventually reaches a derelict museum, full of the usual exhibits and displays, where they find, hidden under a curtain, the sought-after sorrow gate mother: a techno-mammoth made from "steel and spark," now rusted and crusted over with moss, forged long ago to teach the de-extincted flesh mammoth how to *be* a mammoth. Her songs are needed once more to heal the mammoth, and for a while, the mamontogist is able to revive her, talk with her, learn from her. But when it comes to making the journey back to the tribe, she is unable to summon enough power to stand; the narrator is forced to leave her behind in Plastic Park and return home alone. Once there, the narrator implores the rest of the group to travel back to the sorrow gate mother with them to fix her, and declares that they shall never "dumbeasticate" her as Plastic Park had done.

Chrulew's story stands in almost direct opposition to Obruchev's *Plutonia,* in that it offers a much bleaker but also more hopeful future. In "The Mamontogist's Tale," the human species appears to have been decimated, and the reader is left in no doubt that this was due to anthropogenic destruction and hubris, much of which is embodied in the rendering of the Pleistocene/Plastic Park as a gaudy tourist attraction in the vein of its obvious comparison, Jurassic Park. Yet despite the emptiness of the park's paddocks and fields, the dying of the mammoths, and the rusting over of the sorrow gate mother, the park offers a spark of hope in the sprawling tundra amid the undergrowth. The techno-mammoth survives! No longer forced to perform for her human masters, bound by fences, buttressed by capital, her almost shattered form can be made anew, her re-resurrection can make her *world* anew alongside other forms of life and nonlife. The point is not to advocate for some primitivist return to a previous time, and certainly not to advocate for the demise of humanity. Rather, what Chrulew's story offers, and what I believe the thawing of permafrost can also offer, is the chance to imagine new worlds that are not bound to the yoke of capital or techno-utopic fantasy. This is a future-gazing story that opens up the possibilities of reclamation rather than appropriation, resistance rather than erasure, an emergent ruination rather than degradation.

As with the bones that litter Duvanny Yar and produce temporal frictions amid ecological ruins, so might the remains of de-extincted mammoths reverberate across future landscapes and timescales. This is the generative potential of extinction, its ability to hold the awkward categories of absence and presence in tandem. What Chrulew's story also hints at is that any future reimagining must also be decolonial, and by redefining apocalypse and extinction as saturated with coloniality, it allows for a much more expansive and historic understanding of where and how these processes occur. Whoever the protagonists of "The Mamontogist's Tale" are, what we do find out about them is that they worship the mammoth. Might their ancestors have been the Evenki, or the Yukaghir? Might their stories and cosmologies echo those that feature the mammoth as an underground god? Permafrost thaw lays bare the long arm of colonialism in Siberia and the apocalypses that have already happened for many Indigenous groups worldwide, yet in its becoming discontinuous it pushes us to pay attention to the visible holes in the dominant Western ontological positioning. Permafrost natures across time and space have *always* been discontinuous, but the cracks that are forming *now* reveal a sense of urgency, not only to think heterogeneously beyond species and ontological boundaries but also to understand that decoloniality is a long process of resistance and remaking with earth.[19] Through a more expansive and hopeful reimagining of the de-extinction trajectory that currently exists, extinction itself also becomes, or *can* become, something more.

In a short story that imagines a postapocalyptic interracial love relationship that is possible only because all the other white people are dead—a speculative fiction that demands the extinction of whiteness as an ideology rather than the deaths of millions—Saidiya Hartman points exactly to the sort of largeness the idea of extinction might have if opened up beyond its normative definition. She writes: "The stranglehold of white supremacy appears so unconquerable, so eternal that its only certain defeat is the end of the world, the death of Man."[20] Throughout this book, I have been probing at this idea—that discontinuity encompasses the destructive potential of anthropogenic climate change, but also perhaps the discontinuation of the destructive practices that have brought about the Anthropocene in the first place.

What the coming years will bring cannot be predicted with any degree of certainty. Whatever happens at the Pleistocene Park will depend on multiple factors, most of which are uncontrollable. The global Covid-19 pandemic thwarted researchers hoping to visit NESS for two summers, in 2020 and 2021—a huge blow to the park's finances. Nikita's new expedition to Wrangel Island to find some more musk oxen stalled at the first hurdle, with the little boat finding itself caught in a storm and unable to go any farther. Now, Russia's invasion of Ukraine means that access to the park is almost impossible for non-Russians. The project to save the world has, for now, been put on hold. Meanwhile, in Yakutsk, the mammoth de-extinction project was rocked by the untimely death of Semyon Grigoriev, head of the Mammoth Museum. The permafrost continues to thaw, however, and temperatures in the Arctic during the summer of 2020 reached a staggering 38 degrees Celsius. Chersky was surrounded by wildfires; permafrost disintegration caused extensive structural damage in the town, including the collapse of an entire housing block; and a fuel tank in Norilsk spilled vast quantities of diesel into the river, staining the water blood red as the fuel snaked its way down to the Arctic Ocean.[21]

"The challenge of decolonizing extinction, then," states Juno Salazar Parreñas, "is not to end extinction, but to consider how else might it unfold for those who will perish and for those who will survive."[22] This must be the point of thinking discontinuously. Throughout this book I have argued for a reimagining of the normative definition of extinction and a dismantling of the Western orientation of the planetary—a world of white supremacy, colonialism and empire, heteropatriarchy, and scientific singularity. This is not a world worth saving. We are apparently now a planetary species as Anthropos, but without an interrogation into what this means, the definition is a shallow one—a surface, a shiny globe. What is beneath and what is under matters a great deal when it comes to recognizing the multiplicity of relations with the Earth. How we inhabit our own small worlds yet are also part of a planetary collective is a discontinuous, but nevertheless entangled, identity. I do not offer discontinuity as a panacea, but rather as a point of departure for imagining new worlds that refuse the dominant paradigm. My own imagination stretches only so far, so I call on others to imagine as well. This will mean different interpretations and different orientations; sometimes, as with permafrost, they may remain hidden or in isolation. They will never match up or form a complete whole, nor should this be the aim, but there is scope for imagining the ways in which they can be held together in solidarity, in strength and fragility, on a discontinuous Earth.

Acknowledgments

Between the time when I completed my PhD in 2020 and the publication of this book, it is fair to say a lot has happened, including a global pandemic that is still raging and a needless, brutal war on Ukraine instigated by Russia, which is now the recipient of the largest package of international sanctions in history. The world is not the same as it was before all this, and it will never be the same again. Because I have written this book in and about Russia, some of it is now hopelessly out of date—an anachronistic glance back to a time that seems like a hundred years ago but is actually less than five. But among the stories of a time when international research and collaboration with Russian scientists, not to mention travel to and within Russia, were possible, there is a continuous thread: the permafrost is thawing as a result of climate change and continues to thaw regardless of war and plague. I hope that Arctic research can continue across borders and that the Russian scholars working on the Arctic are not excluded from collaborative conversations. This book is situated in Russia, but what happens in the Arctic bleeds beyond the boundaries of nation-states, outside the remit of political regimes. The people found in the pages of this book have voices that deserve to be heard—not closed off from the rest of the world. My thanks are to them above all.

Special thanks go to the people who made this research possible in Russia: Alexander Fedorov at the Melnikov Permafrost Institute in Yakutsk, Semyon Grigoriev at the Mammoth Museum in Yakutsk (sadly, Semyon died in 2020), and, of course, the Zimov family in Chersky. Thank you to Galya for cutting through the behemoth of Russian bureaucracy, to Sergey for his brutal honesty, to Nastya for her organizational skills and friendship, and to Nikita for being so candid and thoughtful—this book would be a poor shadow without his words. Whether the Pleistocene Park will survive in a Russia shut off to the international researchers who fund it remains to be seen, but I fear the worst. Nikita, with his relentless stoicism, merely told me he has no other option than to keep going.

My thanks to the friends I made along this journey, who have enriched my scholarship immeasurably and provided much-needed fun and support in equal measure: Theo Barry-Born, Rosalie Warnock, Lucie Glasheen, Faith Taylor, Shereen Fernandez, Vincent Guermond, Teresa Aguilar, Ben Halligan, Aurora Fredriksen, Ben Garlick, Earl Harper, Justin Pickard, Alex Damianos, Adam Searle, Jonny Turnbull, Sarah Bezan, Sarah Pickman, Miranda Cichy, Yulia Zaika, Ingrid Medby, and Mia Bennett. And of course, the (Anthropo)cene crew: Mariana Reyes, Matthew Beach, Katy Lewis Hood, and Therese Keogh.

Thank you to Sam Saville and Jen Telesca, the two readers whose thoughtful and insightful comments improved this book immeasurably.

I am grateful also to my academic mentors over the years: Nick Higham, Franklin Ginn, Kathryn Yusoff, Catherine Nash, Julia Lajus, and now Dolly Jørgensen. This book would not exist without their support, encouragement, and patience, as well as their scholarship, from which I took so much inspiration.

Thank you to everyone at the University of Minnesota Press, who have been a joy to work with and were particularly supportive during a time of great upheaval in my life. Special thanks to my editor Jason Weidemann, who took the time to listen to my pitch at the AAG meeting in 2019—and has been with me every step of the way since!

Thank you to my nonacademic friends who have supported my work (some even helped me proofread), who listened when they weren't sure what was going on, and continue to be endless sources of love: Lesley, Sally, Charlee, Graeme, Margo, Elena, Emily, Mel, Ying, Luke, Jo, Matt, Jen, and Mystery Club—Pokes, JJ, Weevs, Marc, Silvs, and Bob. And to my family, thank you for everything.

I write this two weeks into Russia's war on Ukraine, barely able to tear myself away from the rolling news of horror. I condemn this brutal attack utterly, but I can't help but feel a great sadness for the good, kind, peaceful Russians I know. I lived and worked in Russia for more than a year, and I left very quickly like a thief in the night, on a plane full of Russians fleeing to Istanbul. This breaks my heart, not just for what I have lost but also for what many Russians who did not ask for this senseless war will lose. Their futures are now bleak. I dedicate this book to everyone who enriched my life throughout fifteen years of visiting Russia, especially Kostya, Eve, Denis, Anna, Aynur, Dasha, Ruslan, Slava, Yulia, Yana, Sonia, Fyodor, Alexey, Tanya, Alex, Roman, Anna, Julia, and Nikolay—and others I'm sure I'm forgetting. And to Lyosh, whom I will miss most of all. Courage, my friends. I know we will meet again.

Notes

Preface

1. Chu, *The Life of Permafrost.*
2. Ingold, *The Perception of the Environment,* 153.
3. Geertz, *The Interpretation of Cultures.*
4. Povinelli, *The Empire of Love,* 7.
5. Cohen, *Stone,* 9.
6. Sörlin, "Cryo-history."

Introduction

1. Chu, *The Life of Permafrost.*
2. Chu, *The Life of Permafrost.*
3. Dobinski, "Permafrost."
4. International Permafrost Association, "What Is Permafrost?"
5. Turetsky et al., "Permafrost Collapse." Subsea permafrost also exists in the Arctic Ocean, but for the purposes of this study, only land permafrost is considered.
6. Yedoma is carbon-rich permafrost with a high ice content that makes it vulnerable to rapid thaw and the concurrent release of greenhouse gases. Sakha, in particular, has an abundance of yedoma soils.
7. IPCC, *Special Report on the Ocean and Cryosphere.*
8. IPCC, *Global Warming of 1.5°C.*
9. Tatarchenko, "The Lena Is Worthy of Baikal."
10. The construction of the Trans-Siberian Railway, begun in 1891 and completed in 1904, allowed for the transportation of further materials for infrastructure development, and, of course, Stalin's brutal regime of penal labor saw gulags emerge in some of the most inhospitable parts of the country. Josephson, *The Conquest of the Russian Arctic.* Without this huge mobilization of forced servitude and torture, much of the built environment in these difficult spaces would never have been realized. Entire towns and other infrastructure, such as mines and factories, were built on the backs of gulag laborers, as illustrated by Magadan's Road of Bones, so named because of the huge death toll among the prisoners who built it. Taking place alongside this rapid expansion of new settlements and work opportunities was the movement of Russians

eastward. As Peter Schweitzer and his colleagues note: "Building socialism implied not only an accumulation of wealth, but also of people." Schweitzer at al., "Beyond Wilderness," 66.

11. Bolotova, "Colonization of Nature."
12. Heleniak et al., "Cities on Ice"; Premiyak, "The Last Frontier."
13. Liu and Kronbak, "The Potential Economic Viability." The new Arctic strategy sets out a plan to quell the population exodus through a series of incentives to stay, including the creation of jobs and social enterprises, while admitting much of this funding would have to come from private investment. Covid-19 and the collapse in oil prices have caused further uncertainty regarding the implementation of the strategy, and there is little mention of accompanying Arctic infrastructure such as roads and power lines—many of which are in a state of disrepair. Aliyev, "Development in Difficult Times."
14. Laruelle, "Russia's Arctic Policy."
15. Crutzen and Stoermer, "The Anthropocene"; Zalasiewicz et al., "Are We Now Living in the Anthropocene?"
16. Ceballos et al., "Accelerated Modern Human-Induced Species Loss."
17. Foucault, *The History of Sexuality*; Radin and Kowal, "Introduction," 6.
18. Watt-Cloutier, *The Right to Be Cold.*
19. Blinnikov et al., "Pleistocene Graminoid-Dominated Ecosystems."
20. The full territory owned by the park is approximately 144 square kilometers, which was gifted to the Zimovs by the Sakhan government (a shock to Sergey and Nikita, given how little support they had received previously). While the park remains in its experimental stage, however, the fenced smaller area is necessary to allow for monitoring of the animals.
21. Krupnik and Jolly, *The Earth Is Faster Now. Alaas* are formed from thermokarst that happens slowly, over hundreds of years; the lakes that form in the thermokarst hollows eventually drain away, leaving behind lush fields of nutrient-rich grasses and plants. These little tundra oases are very sensitive to climate, and more rapid thermokarst has a detrimental effect on their formation. Crate et al., "Permafrost Livelihoods."
22. Luhn, "Anthrax Outbreak."
23. On the overkill argument, see Zimov, "Pleistocene Park." On the hypothesis that climate caused the changes, see Levy, *Once and Future Giants.*
24. Much of the promotional material for the park, aimed at encouraging donations, is curated along these "all or nothing," world-saving lines. The project's first crowdfunding campaign ran with the headline "An Ice Age Ecosystem to Save the World." Pleistocene Park Foundation. "Pleistocene Park."
25. Cohen, *The Fate of the Mammoth.*
26. Weiss, "The Climate Crisis."
27. Falling into a third category, sporadic, are those small pockets of perma-

frost found in latitudes normally not cold enough for the ground to freeze but where other variables (altitude, shade) may keep the temperature low enough for permafrost to form.

28. International Permafrost Association, "What Is Permafrost?"
29. Romanovsky et al., "Thermal State of Permafrost"; Welch, "Some Arctic Ground."
30. Wolfe, "Foreword."
31. Grusin, "Introduction"; Parreñas, *Decolonizing Extinction.*
32. Colebrook, *Death of the Posthuman.*
33. Red lists, which have been critiqued by a number of scholars, are problematic in that their categorizations rely heavily on species markers. See Heise, *Sense of Place*; Braverman, "En-listing Life."
34. Van Dooren, *Flight Ways,* 12.
35. Rose, "Multispecies Knots of Ethical Time." On humans' responsibility to multispecies kin, see Haraway, *Staying with the Trouble.*
36. Wolfe, "Foreword," viii.
37. Chakrabarty, "The Climate of History"; Haraway, *When Species Meet.*
38. Moynihan, *X-Risk.*
39. Grusin, "Introduction."
40. Balmford, *Wild Hope*; Avery, *A Message from Martha.*
41. Heise, *Imagining Extinction;* Turner, "Open-Ended Stories."
42. Garlick and Symons, "Geographies of Extinction."
43. Moynihan, *X-Risk,* 12.
44. Grove, "Of An Apocalyptic Tone"; Northcott, "Eschatology in the Anthropocene."
45. Heise, *Sense of Place.*
46. Brand, *Whole Earth Discipline.*
47. Sepkoski, *Catastrophic Thinking.*
48. Pugh and Chandler, *Anthropocene Islands.*
49. IUCN, "IUCN Red List of Threatened Species."
50. Chakrabarty, "The Climate of History"; Chakrabarty, "Critical Zones."
51. Malm and Hornborg, "The Geology of Mankind?"
52. Crist, "On the Poverty of Our Nomenclature"; Moore, "Anthropocene or Capitalocene?"; Nixon, "The Anthropocene"; Tsing, *The Mushroom at the End of the World.*
53. Colebrook, "The Future in the Anthropocene."
54. Grusin, "Introduction," vii.
55. Davis and Todd, "On the Importance of a Date"; DeLoughrey, *Allegories of the Anthropocene.*
56. Milkoreit, "Imaginary Politics."
57. Ginn, "When Horses Won't Eat"; Yusoff and Gabrys, "Climate Change and the Imagination."
58. Danowski and Viveiros de Castro, *The Ends of the World;* Szerszynski, "The End of the End of Nature."
59. Northcott, "Eschatology in the Anthropocene."

60. Chu, "Mapping Permafrost Country"; Chu, *The Life of Permafrost.*
61. Radin and Kowal, "Introduction," 13.
62. Dodds, "Geopolitics and Ice Humanities."
63. Grosz et al., "An Interview"; Povinelli, *Geontologies.*
64. Povinelli, *Geontologies.*
65. Yusoff, "Geologic Subjects," 388.
66. Carey, "The History of Ice."
67. The ecologist Frans Vera has written about a "shifting baseline syndrome" (a term coined by Daniel Pauly in 1995) in restoration conservation, in which the ecological conditions of a landscape are informed by human memory. Vera, "The Shifting Baseline Syndrome."
68. Cruikshank, *Do Glaciers Listen?*
69. Parreñas, *Decolonizing Extinction.*
70. Schmidt, "Glacial Deaths."
71. WWF, *Living Planet Report.*
72. Malm and Hornborg, "The Geology of Mankind?"; Collard and Dempsey, "Capitalist Nature in Five Orientations"; Wrigley, "Ice and Ivory."
73. Telesca, *Red Gold.*
74. Mitchell, "Decolonizing against Extinction."
75. Braverman, "En-listing Life."
76. Van Dooren, *Flight Ways*; van Dooren and Rose, "Keeping Faith with the Dead"; Haraway, *When Species Meet.*
77. Wolfe, "Foreword."
78. Jørgensen, "Endling."
79. Garlick, "Cultural Geographies of Extinction"; Garlick and Symons, "Geographies of Extinction"; Wrigley, "Nine Lives Down."
80. Zimov et al., "Mammoth Steppe."
81. For arguments in favor of rewilding, see Monbiot, *Feral*; Bekoff, *Rewilding Our Hearts.* On the other side, see Jørgensen, "Rethinking Rewilding." For my own concerns, see Wrigley, "Nine Lives Down."
82. Garlick, "Cultural Geographies of Extinction," 226.
83. Van Dooren, *Flight Ways.*
84. Lorimer, "Nonhuman Charisma."
85. Colossal, "De-extinction."
86. Yamagata et al., "Signs of Biological Activities."
87. Cohen, "The Ethics of De-extinction"; Richmond et al., "The Potential and Pitfalls of De-extinction"; Sandler, "The Ethics of Reviving Long Extinct Species."
88. Cohen, "The Ethics of De-extinction."
89. Telesca, *Red Gold.*
90. Rockström et al., "A Safe Operating Space for Humanity."
91. Bennett et al., "Bright Spots."
92. Brand, "We Are as Gods."
93. Church and Regis, *Regenesis,* 127.
94. Searle, "Anabiosis and the Liminal Geographies of De/extinction," 331.

95. Rose, "The Politics of Life Itself"; Rose, *The Politics of Life Itself*; Franklin, "Life Itself." Discussion pops up occasionally about the possibility of eradicating malaria through the extinction of mosquitoes (or at least the destruction of the mosquito species that carry malaria). Neslen, "US Military Agency Invests $100m." As much as mosquitoes made my time in Arctic Siberia pretty hellish, I would argue that this is an ethical decision that no one species should be allowed to make.
96. Adams and McCorristine, "Ghost Species."
97. Rose, "In the Shadow of All This Death."
98. Ingold, "A Northern Ontology?"
99. Sörlin, "Cryo-history," 327.
100. I am inspired particularly by Siberian Indigenous scholars and curators, some of whom I met during my visits to the North-Eastern Federal University and some in Chersky, who do not all have written or accessible bodies of work. These perhaps more intangible perspectives are difficult to grasp and recount on the page, so inspiration here happens in a more abstract sense, as a way to point to cosmologies that do not fit the Western academic model.
101. Cosgrove and Della Dora, "Introduction."
102. Ingold, *Hunters, Pastoralists and Ranchers*; Ingold, *The Perception of the Environment.*
103. Barad, *Meeting the Universe Halfway*; Bennett, *Vibrant Matter*; Braun and Whatmore, "The Stuff of Politics"; Coole and Frost, "Introducing the New Materialisms"; Cohen, *Stone.*
104. Cameron, *Far Off Metal River.*
105. Swyngedouw, "Apocalypse Now!"

1. Earth

1. Thermokarst is a geophysical process whereby hollows appear in the permafrost due to thaw; these hollows often then fill with water to create lakes. See Lara et al., "Thermokarst Rates Intensify."
2. Haraway, "Anthropocene, Capitalocene"; Moore, "Anthropocene or Capitalocene?"; Tsing, *The Mushroom at the End of the World.*
3. Haraway et al., "Anthropologists Are Talking."
4. Farnham, *Saving Nature's Legacy.*
5. Elden, "Land, Terrain, Territory."
6. Elden, "Terrain, Politics, History," 180.
7. Gordillo, "The Power of Terrain," 191.
8. Aradau and van Munster, *Politics of Catastrophe*; de Goede and Randalls, "Precaution, Preemption"; Methmann and Rothe, "Politics for the Day after Tomorrow"; Mitchell et al., *Global Environmental Assessments.*
9. Natali et al., "Permafrost Carbon Feedbacks."
10. Quoted in Welch, "Some Arctic Ground No Longer Freezing."
11. Fedorov et al., "Permafrost-Landscape Map."

12. Methmann and Rothe, "Politics for the Day after Tomorrow," 330.
13. PCF Action Group, "Are Permafrost Thaw Interventions Possible?"
14. National Research Council of Canada, *Glossary of Permafrost.*
15. A selection of popular science articles on permafrost thaw published in 2018 adopt a tone that conveys a vague sense of creeping dread. Statements such as "The stakes are high," "Something stirs: what will happen when permafrost thaws?," and "The consequences of climate change can be weird and apocalyptic" speak to this uncertainty in a dramatic fashion. Welch, "Some Arctic Ground No Longer Freezing"; Alex, "Something Stirs"; Resnick, "Melting Permafrost in the Arctic." In order to lend a certain amount of evidence and expertise to the articles, these vagaries are punctuated by quotes from scientists who, loath to commit to any startling conclusions while not wishing to underplay the severity of the situation, offer insights such as "If you assume it's a trend or that it might continue like this, then it's alarming" (quoted in Welch) and "The more we push the system, the less control we have" (quoted in Nash, "The Impermanence of Permafrost").
16. Meyer, "The Zombie Diseases of Climate Change."
17. Colebrook, "The Future in the Anthropocene," 264.
18. Iida, "Fukushima."
19. Grusin, "Introduction."
20. Moynihan, *X-Risk,* 31.
21. Livingstone, *Putting Science in Its Place;* Mitchell et al., *Global Environmental Assessments.*
22. Loranty et al., "Understory Vegetation."
23. Alexander et al., "Impacts of Increased Soil Burn Severity."
24. Cormier, "Why the Arctic Is Smouldering."
25. See Nakano et al., "Temporal Variation in Methane Emission"; Corradi at al., "Carbon Dioxide and Methane Exchange"; Dutta et al., "Potential Carbon Release."
26. Geissler and Kelly, "A Home for Science."
27. David Livingstone deals with this phenomenon well in *Putting Science in Its Place,* noting that while amateur scientists are able to access the field, if they cannot access the more legitimate space of the laboratory, where professional science is done, their findings are not taken seriously. The coming together of different knowledges is made possible by the dinner table at the science station, but the science done in the field is rendered legitimate only when the scientists take it back to their institutions for peer review. The Zimovs occupy a curious in-between space where they can embody the roles of both amateur and professional scientist.
28. Stengers, "Reclaiming Animism."
29. Slater, *Mammoth.*
30. Kelly, "The Man-pocalypse."
31. Heise, *Sense of Place.*

32. Pleistocene Park Foundation, "Pleistocene Park"; Pleistocene Park Foundation, "Bison to Save the World."
33. Swyngedouw, "Apocalypse Forever?"; Swyngedouw, "Apocalypse Now!"
34. McCannon, *Red Arctic*; McCannon, *A History of the Arctic.*
35. During the Soviet era, charter helicopter flights to and from Chersky were free, or at least heavily discounted. Sergey reminisces one evening that "we used helicopters like this to bring hay for deer, or bring just one movie for reindeer people to show them." Nikita tells me a ludicrous tale of a group of men who flew from Chersky to Yakutsk to buy beer because it was cheaper there than at the local shop.
36. Sukhova et al., *Demograficheskiye protsessy Respublike Sakha.*
37. Arctic residents who have lost their livelihoods are eligible for rehousing by the Russian government, but this process is incredibly slow; many have remained stuck for thirty years. Luhn, "Warming Climate and Arctic Gas Push."
38. Pushkareva and Burykin, *Fol'klor narodov Severa,* 357.
39. DeLoughrey, *Allegories of the Anthropocene.*
40. Pugh and Chandler, *Anthropocene Islands.*
41. Moore, "Anthropocene Anthropology."
42. Holbraad et al., "Introduction."
43. Slezkine, *The House of Government.*
44. Dennis, "A Polar Perspective"; Doel, "Constituting the Cold War Earth Sciences."
45. Holbraad et al., "Introduction," 1.
46. Povinelli et al., "Quasi-events."
47. Colebrook, *Death of the Posthuman,* 70.
48. Drake, "Elon Musk."
49. Smirnov, "The Permafrost." The art exhibition was held in the permafrost tunnel at the Melnikov Permafrost Institute. Smirnov has written extensively around permafrost science, geography, and animism from an artistic perspective, and he exhibited several of my fieldwork photographs at his show on the subterranean at the Ural Biennale in 2019. See also Smirnov, "The Provincialization of the Globe."
50. Spivak, *Death of a Discipline,* 72.
51. Farrier, *Footprints.*
52. The *Olonkho* is a series of (usually sung) Sakhan epic tales, each containing many verses, usually detailing heroic exploits and daring quests of the Sakhan people, and often with a moral message. Very few of these tales have been written down, and traditionally a tribe would have a special narrator skilled in the singing of *Olonkho* from memory—an incredible feat, given that some of the stories are as many as forty thousand verses long, with recitations lasting for eight hours.
53. Luhn, "Anthrax Outbreak."
54. Alexander et al., "Impacts of Increased Soil Burn Severity."
55. Welch, "Some Arctic Ground No Longer Freezing."

56. It is worth pointing out that Sergey and Nikita (the latter to a lesser extent) have curated the role of "outsiders" for themselves, but this is certainly not an "outsider within" situation of the type detailed carefully and thoroughly by Black women in the academy who know they can never truly belong while it remains a bastion of white supremacy (for example, see Collins, "Learning from the Outsider Within"). The Zimovs retain easy access to the "inside" by way of their white male privilege, but they prefer to be seen as rogue scientists pushing back against the stuffy rules of those on the inside—mavericks who get things done *now* rather than track trends for years. In "Apocalypse Soon?" Feinberg and Willer highlight how doom-laden narratives of a climate apocalypse tend to reduce belief in global warming rather than cement it. The role of the institutional expert and the role of the rogue scientist exist in uneasy tandem when it comes to the reporting of climate science; each side is validated—through criticism—by the other. See Winterton, "In Praise of (Some) Mavericks."
57. Welch, "Some Arctic Ground No Longer Freezing."
58. Livingstone, *Putting Science in Its Place.*
59. Stengers, "Reclaiming Animism"; Stengers, *In Catastrophic Times*; Heise, *Sense of Place.*
60. Davis and Todd, "On the Importance of a Date," 774.
61. Davis and Todd, "On the Importance of a Date"; Rice, *Moon of the Crusted Snow*; Whyte, "Indigenous Climate Change Studies"; Whyte, "White Allies."

2. Ice

1. Another reason Yakutsk has no bridge across the Lena is that money earmarked for the project was reportedly rerouted by Putin after the annexation of Crimea in 2014. Bennett, "Putin Opens Bridge to Crimea." To secure Russia's claim to Crimea, Putin ordered a bridge to be built across the Black Sea to link the contested region to mainland Russia. The nineteen-kilometer bridge was built very quickly, and many engineers have warned that its design and construction are not adequate to withstand the seismic activity in the area.
2. The founding of the MPI came right in the middle of what historians call the "Khrushchev thaw," the decade after Joseph Stalin's death when his successor Nikita Khrushchev implemented a rollback of the harsh and "cold" policies that had underpinned Stalin's brutal regime. Many of the esteemed scientists who had been sent to gulags in Siberia because they were considered enemies of the Soviet Union were released and reinstated in their previous positions. Instead of being suppressed, science was celebrated and well funded as part of Khrushchev's concentrated effort to do away with Stalinist excess and renew the Soviet

vision through techno-utopic ideals of progress. Rogacheva, "Soviet Scientists Take the Initiative."

3. Fedorov et al., "Permafrost-Landscape Map."
4. There are twenty-seven different Indigenous groups in Siberia; the main populations who make their home in the Kolyma region are the Even, Chukchi, and Yukaghir, although there are now very few Yukaghir left. All of these groups traditionally practiced nomadic reindeer herding before Soviet cultural assimilation forced them into collective farms, where they were prevented from speaking their own language and practicing their own beliefs. Slezkine, *Arctic Mirrors*. Although the policy of Russification is now over, most of the Indigenous people in Siberia have neither the knowledge nor the desire to return to a nomadic life; those who do face incredibly tough conditions and extremely low pay. For an in-depth study of a group of Eveni who returned to nomadism, see Vitebsky, *Reindeer People*.
5. Haraway, "Situated Knowledges."
6. Kintisch, "These Ice Cellars Fed People."
7. Melnikov Permafrost Institute, "Permafrost Seed Repository."
8. Lena's remarks echo a broader phenomenon in which discussion of weather patterns shifted to discussion of climate patterns as people began to become more aware of planetary conditions. As the terms *climate change* and *global warming* entered modern parlance, coinciding with the environmental movement, which posits all humans as interconnected life-forms on a fragile Earth, weather became subsumed into the larger-scale topic of climate. Hastrup and Skrydstrup, *The Social Life of Climate Change Models*.
9. Barad, *Meeting the Universe Halfway*; Bennett, *Vibrant Matter*; Braun and Whatmore, "The Stuff of Politics"; Coole and Frost, "Introducing the New Materialisms."
10. Grosz, *Becoming Undone*, 31
11. The words in parentheses highlight how permafrost knowledges have always been, to an extent, uncertain and heterogeneous, but large-scale and disruptive shifts are recent phenomena.
12. Salazar, "Ice Cores as Temporal Probes," 77.
13. Leslie, *Liquid Crystals*, 9.
14. Leslie, *Liquid Crystals*; Salazar, "Ice Cores as Temporal Probes."
15. Tarnocai et al., "Soil Organic Carbon Pools"; Lachenbruch and Marshall, "Changing Climate"; Bakermans et al., "Reproduction and Metabolism."
16. I left the park before I learned the fate of these cores.
17. O'Reilly, "Sensing the Ice."
18. Swyngedouw, "Apocalypse Forever?"
19. Republic of Sakha, "Ob okhrane vechnoi merzloty."
20. Quoted in Antonova, "Siberian Region Fights to Preserve Permafrost."
21. Republic of Sakha, "Ob okhrane vechnoi merzloty."

22. Yusoff, "Geologic Life," 780.
23. Quoted in EXO-YKT, "The Ice Has Broken."
24. Quoted in EXO-YKT, "The Ice Has Broken."
25. Povinelli, *Geontologies,* 173.
26. Republic of Sakha, "Ob okhrane vechnoi merzloty."
27. In recent years, in addition to having to contend with permafrost thaw and Arctic heating, the Nenets autonomous oblast, a region rich in natural gas, has endured several large-scale liquefied natural gas projects backed by the Russian government.
28. Gertcyk, "Huge Cull of 250,000 Reindeer."
29. Stépanoff, "Human–Animal 'Joint Commitment.'"
30. Quoted in Arkhangelkskaya, "The Siberian Plague."
31. Heavy snowfalls in Arctic Russia are often accompanied by warmer temperatures, a phenomenon that makes it difficult for the reindeer to feed. In the past decade there have been several heartbreaking instances of rain falling onto snow and then freezing overnight, creating an impenetrable ice crust over the soft snow. When this occurs, the reindeer are unable to break through the ice to get to the moss and grass beneath. If their handlers are not able to provide them with food, they starve to death.
32. Cruikshank, *Do Glaciers Listen?,* 3.
33. Crate et al., "Permafrost Livelihoods."
34. Sleptsov-Sylyk, *Dykhanie vechnoi merzloty,* 128.
35. Sharma, *In the Meantime.*
36. Hird, "Knowing Waste."
37. Parreñas, *Decolonizing Extinction.*
38. For an excellent photo essay on the tusk hunters, see Chapple, "The Mammoth Pirates."
39. Carved ivory ornaments are status symbols in China, and that country now holds the dubious honor of being the world's largest player in the ivory trade. After China placed a ban on elephant ivory in 2018, the Chinese market turned to mammoth ivory to meet the demand. See Myers, "Weaning Itself from Elephant Ivory." While this move may have had a positive impact on elephant poaching elsewhere in the world, the forces of capitalism have made themselves known in other arenas—in this case, the thawing permafrost of Siberia, through which the mammoth's body is renegotiated as a luxury commodity.
40. McKay, *Discovering the Mammoth.*
41. Anderson, "Primal Life Force under the Ice."
42. Grant, *In the Soviet House of Culture*; Cohen, *The Fate of the Mammoth.*
43. Anderson, "Primal Life Force under the Ice."
44. Chapple, "The Mammoth Pirates."
45. Breithoff and Harrison, "From Ark to Bank"; Chrulew, "Freezing the Ark."

46. Tsing, *The Mushroom at the End of the World*; Laruelle, "Russia's Arctic Policy."
47. Josephsen, *The Conquest of the Russian Arctic.*
48. Carey et al., "Glaciers, Gender, and Science."
49. Tsing, *The Mushroom at the End of the World.*
50. McCannon, *Red Arctic.*
51. Tatarchenko, "The Lena Is Worthy of Baikal."
52. Steinberg and Peters, "Wet Ontologies, Fluid Spaces"; Steinberg and Peters, "Cross-Currents and Undertows."
53. Of course, this is not a true absence of matter, as thawing permafrost becomes soil rather than disappearing completely, but it becomes absent in that it is now no longer permafrost.
54. Records of the Ambarchik camp, like those of many Siberian gulags, have either been sealed to all but a select few or destroyed. There are traces of the gulag around Chersky: remnants of possessions displayed as a simple memorial in the museum; places with names like Bloody Lake, which the locals will tell you is haunted; and stories passed down through generations, such as the one about the prisoner ship *Dzhurma* becoming trapped in the sea ice near Ambarchik in 1933, and all twelve thousand people on board perishing. The *Dzhurma* story has now been discounted as unlikely, but the legend endures around town. Bollinger, *Stalin's Slave Ships.*
55. Yusoff, *A Billion Black Anthropocenes.*
56. China, in particular, is prospecting swaths of Greenland as sites for mining uranium and zinc, while positioning itself as a near-Arctic state with its new Belt and Road Initiative, demonstrating that it has serious designs on developing the space for its own interests.
57. Yusoff, *A Billion Black Anthropocenes.*
58. Cruikshank, *Do Glaciers Listen?*

3. Bone

1. Zimov et al., "Mammoth Steppe."
2. Vasil'chuk et al., "First Direct Dating"; Murton et al., "Preliminary Paleoenvironmental Analysis"; Murton et al., "Paleoenvironmental Interpretation of Yedoma Silt"; Yashina et al., "Regeneration of Whole Fertile Plants"; Weisberger, "Worms Frozen for 42,000 Years." I say "allegedly" regarding the worms' age because this is a very spurious claim—most likely the worms were much younger.
3. Zimov, "Pleistocene Park"; Zimov et al., "Mammoth Steppe."
4. Squire and Dodds, "Introduction."
5. Froese et al., "Ancient Permafrost."
6. Hoag, "Permafrost That Lives Up to Its Name."
7. Clark, *Inhuman Nature.*

8. Melo Zurita et al., "Un-earthing the Subterranean Anthropocene"; Squire and Dodds, "Introduction."
9. Jules Verne's *Journey to the Center of the Earth* (1864) springs to mind, as well as its Russian counterpart, *Plutonia,* by Vladimir Obruchev (1924). J. R. R. Tolkien's *Lord of the Rings* (1954) depicts the follies of mining: by digging "too greedily and too deep," the dwarves allow an ancient demon to emerge.
10. Kearnes and Rickards, "Earthly Graves for Environmental Futures," 53.
11. Clark, "Politics of Strata," 214.
12. Rudwick, *Bursting the Limits of Time.*
13. Rudwick, *Earth's Deep History,* 80.
14. Sepkoski, *Catastrophic Thinking.*
15. Harries, "Of Bleeding Skulls"; Krmpotich et al., "The Substance of Bones."
16. Verdery, *The Political Lives of Dead Bodies,* 27.
17. DeSilvey and Edensor "Reckoning with Ruins"; DeSilvey, *Curated Decay.*
18. Adams and McCorristine, "Ghost Species."
19. Stoler, " 'The Rot Remains,' " 9.
20. Gordon, *Ghostly Matters.*
21. DeSilvey and Edensor, "Reckoning with Ruins."
22. Zalasiewicz et al., "Stratigraphy of the Anthropocene"; Zalasiewicz et al., "Human Bioturbation."
23. Olson and Messeri, "Beyond the Anthropocene," 29.
24. Sample, "Oldest *Homo sapiens* Bones Ever Found."
25. Zimov et al., "Steppe–Tundra Transition."
26. The "climate versus humans" Pleistocene extinction debate began around the time when Paul Martin introduced his theory of Pleistocene overkill in the 1960s, positing that humans systematically destroyed the megafauna of the Pleistocene era through hunting—simply because they could. This theory remains controversial to this day; as naturalist Adrian Lister put it in 2005, "Even if [humans] had the capacity to kill that many animals, why would they have needed to?" Quoted in " 'Pleistocene Park' Experiment." Many scientists point to climatic shifts as the reason for the high rate of Pleistocene megafaunal extinction. For a good overview of the debate, see Levy, *Once and Future Giants.*
27. Wrigley, "Nine Lives Down."
28. Jackson, "Glaciers and Climate Change."
29. Farrier, *Anthropocene Poetics.*
30. Jørgensen, *Recovering Lost Species.*
31. Jørgensen, "Rethinking Rewilding."
32. Vernadsky, "The Biosphere and the Noösphere."
33. Rispoli and Grinevald, "Vladimir Vernadsky"; Guillame, "Vernadsky's Philosophical Legacy."
34. Vernadsky, "The Biosphere and the Noösphere," 9.

35. Bailes, "Soviet Science in the Stalin Period."
36. Zimov, "Wild Field Manifesto."
37. Zimov, "Wild Field Manifesto," 10. This is an ambiguous turn of phrase, but Sergey is referring to the fact that (some) humans see themselves as "Earthmasters," rather than suggesting that this is a goal humans should strive toward. See Hamilton, *Earthmasters.* Sergey's personal philosophy is a little more complicated, as his Pleistocene Park project certainly advocates for benevolent planetary stewardship in the vein of Vernadsky, but he also talks of the human species as part of an ecological collective rather than a separate category.
38. Carey, "The History of Ice."
39. The bison made it to the park safe and sound in September 2019 and are now reportedly thriving.
40. Bekoff, *Rewilding Our Hearts*; Monbiot, *Feral.*
41. Bastian, "Fatally Confused."
42. Bear and Holloway, "Beyond Resistance"; Whatmore, "Materialist Returns."
43. Lorimer and Driessen, "Bovine Biopolitics"; Lorimer and Driessen, "Wild Experiments"; Lorimer and Driessen, "From 'Nazi Cows.' "
44. Adams and McCorristine, "Ghost Species."
45. Wenzel, *Bulletproof,* 3.
46. Salih and Corry, "Displacing the Anthropocene."
47. Gordillo, "The Void," 236.
48. Wenzel, *Bulletproof.*
49. Rockström et al., "A Safe Operating Space for Humanity."
50. Pearson, "Beyond 'Resistance.' "
51. Wynne-Jones et al., "Feral Political Ecologies?"
52. Garlick and Symons, "Geographies of Extinction"; Wolfe, "Foreword."
53. Salih and Corry, "Displacing the Anthropocene."
54. Hamilton, *Earthmasters.*
55. Gordillo, "The Void."
56. Jackson, "Glaciers and Climate Change."
57. Stengers, "The Challenge of Ontological Politics," 108.

4. Blood

1. Scanlon, "Korea's National Shock at Scandal."
2. Palkopoulou et al., "Complete Genomes Reveal Signatures."
3. Sussberg and Alvarado, *We Are as Gods.*
4. Colossal, "The Mammoth."
5. "Scientists 'Confident' That They Can Extract Cells."
6. The only "successful" de-extinction attempt so far, the cloning of a Pyrenean ibex (bucardo), saw the creature live for just ten minutes because of a congenital lung defect. For a thorough interrogation of the

bucardo de-extinction efforts, see Searle, "Anabiosis and the Liminal Geographies of De/extinction."

7. I have mentioned this previously, but it is worth repeating that Semyon died in April 2020. What will happen with the partnership between the Mammoth Museum and Sooam Biotech, and their de-extinction efforts, is now unclear.
8. Palmer, *Fossil Revolution.*
9. Rudwick, *Earth's Deep History.*
10. Brook and Bowman, "The Uncertain Blitzkrieg."
11. Fletcher, *De-extinction and the Genomics Revolution.*
12. Deniau, *Raising the Mammoth.*
13. Powell, *Woolly Mammoth.*
14. I purposely gender the "Siberian mammoth" as female, partly because the mammoth body that is most important to de-extinction research (Buttercup) is female, and partly as a way to point to the masculine nature of de-extinction science and the task of the first de-extincted mammoth to begin a cycle of reproduction and birth.
15. This incident was captured on video and can be viewed online: "Russians Find Mammoth Carcass with Liquid Blood," YouTube, posted May 30, 2013, https://www.youtube.com/watch?v=YKPSfok8a7E.
16. Carsten, "Introduction," 2.
17. Carsten, "Introduction," 2.
18. Mammoth tusks provide incredible records of the creatures' lives. With rings like trees trunks, a tusk can reveal a mammoth's age, how many young she bore, whether they reached maturity, and when she entered menopause. Powell, *Woolly Mammoth.*
19. Or, indeed, is it still *de*-extinction? The Long Now Foundation's Ben Novak argues that cloning from cells taken from live creatures (as was the case with the Pyrenean ibex) does not represent true de-extinction, as the cells continued "living" even though the last individual had died. Novak, "De-extinction."
20. Radin, *Life on Ice.*
21. Bruyere, "Puss-in-Boots Goes to Pleistocene Park."
22. Cohen, "The Ethics of De-Extinction."
23. In one pertinent example, the discovery of a well-preserved wolf's head in a permafrost cave went viral on social media. Solly, "A Perfectly Preserved 32,000-Year-Old Wolf Head."
24. Breithoff and Harrison, "From Ark to Bank."
25. Palkopoulou et al., "Complete Genomes Reveal Signatures"; Rajan, *Biocapital.*
26. Other cryobanks are involved in more applied and hands-on science, of course. The San Diego Frozen Zoo, for example, actively participates in endangered species conservation through artificial insemination and stem cell embryonic research; one of its main projects is rescuing the

northern white rhinoceros from the brink of extinction. The critically endangered northern white rhino is actually a subspecies of the white rhinoceros, and its southern counterpart is not at all endangered. This raises questions about the specific value of taxonomy and the performativity of classification systems: If the northern white rhino dies out, is that actually an extinction? This is a particular extinction narrative that has been mobilized by media coverage of the San Diego Frozen Zoo project, while creatures that are deemed less newsworthy go extinct without notice. For more information on the project, see San Diego Zoo Wildlife Alliance, "White Rhino."

27. Landecker, *Culturing Life*, 176.
28. Chrulew, "Freezing the Ark."
29. Mendes, "Molecular Colonialism"; Rose, "The Politics of Life Itself"; Rose, *The Politics of Life Itself*; Thacker, *The Global Genome*.
30. Adams, "Geographies of Conservation I."
31. Van Dooren, "Banking Seed."
32. Powell, *Ice Age*.
33. Searle, "Anabiosis and the Liminal Geographies of De/extinction," 339.
34. Powell, *Woolly Mammoth*.
35. Worrall, "We Could Resurrect the Woolly Mammoth."
36. Waldby, *The Visible Human Project*.
37. Farman, "Cryonic Suspension," 311.
38. Radin, *Life on Ice*.
39. Braun, "Biopolitics and the Molecularization of Life."
40. Landecker, *Culturing Life*.
41. The early twentieth century saw the writings of an ascetic librarian named Nikolay Fedorov become popular with certain circles of the intelligentsia in Moscow. In a philosophy retroactively dubbed "cosmism," Fedorov combined a sort of Christian mysticism with what he described as "the common task"—namely, the duty of all humans to achieve immortality through organic bodily augmentation, and to resurrect their ancestors in order to create utopia. With the advent of the Bolshevik Revolution in 1917, cosmism in its various guises fit rather well with the techno-utopic ideals of the Marxist-Leninist Soviet Union, and it has managed to hang on to the fringes of Russian academic society to this day. Russia is currently the only Eurasian country that has an active human cryobank, where a patron can pay $30,000 to freeze his or her body after death, or $10,000 to freeze only the brain. Bernstein, *The Future of Immortality*.
42. Chu, "Mapping Permafrost Country"; Chu, *The Life of Permafrost*.
43. Sumgin, Vechnaya merzlota pochvy v predelakh SSSR, 15.
44. Sumgin, Vechnaya merzlota pochvy v predelakh SSSR, 354.
45. Bravo, "A Cryopolitics to Reclaim Our Frozen Material States."
46. Sumgin, Vechnaya merzlota pochvy v predelakh SSSR, 351.

47. There is also the practice of "hunting" for extinct species on the off-(and thrilling) chance that experts might be wrong. Adams and McCoristine, "Ghost Species." The 2011 film *The Hunter*, directed by Daniel Nettheim, is a fictional account of one man's hunt for the last thylacine (or "Tasmanian tiger") in the Australian bush to extract its DNA.
48. Farman, "Cryonic Suspension"; Farman, *On Not Dying.*
49. Colossal, "The Mammoth."
50. Fletcher, *De-Extinction and the Genomics Revolution.*
51. Brand, "We Are as Gods."
52. De la Cadena and Blaser, "Pluriverse."
53. Wray, *Rise of the Necrofauna.*
54. Griggs, "Gene Editing."
55. Haraway, "Situated Knowledges."
56. Bruyere, "Puss-in-Boots Goes to Pleistocene Park," 127.
57. Haraway, *Staying with the Trouble*; Kelly, "The Man-pocalypse."
58. Franklin, "From Blood to Genes," 295.
59. Thompson, *Making Parents.*
60. Edelman, *No Future.*
61. Federici, *Re-enchanting the World.*
62. Giffney and Hird, "Introduction."
63. Slater, *Mammoth.*
64. Remarks made in Lohr, *Siberia.*
65. Searle, "Anabiosis and the Liminal Geographies of De/extinction."
66. Farman, "Cryonic Suspension"; Farman, *On Not Dying.*
67. Rose, "Reflections on the Zone of the Incomplete."
68. See Intelligence Squared U.S., "Don't Bring Extinct Creatures Back to Life."
69. Cunha, "The Geology of the Ruling Class?," 263.

Conclusion

1. Although it was written in 1915, *Plutonia* was not published until 1924. The first English translation of the novel did not appear until 1955.
2. Forsyth, *A History of the Peoples of Siberia*; Slezkine, *Arctic Mirrors.*
3. Sleptsov-Sylyk, *Dykhanie vechnoi merzloty.*
4. Remarks in PCF Action Group, "Are Permafrost Thaw Interventions Possible?"
5. Povinelli, *Geontologies*; Grove, "Of an Apocalyptic Tone."
6. Farrier, *Anthropocene Poetics.*
7. Milkoreit, "Imaginary Politics"; Sepkoski, *Catastrophic Thinking.*
8. Putin, "O strategii razvitiya Arkticheskoy zony."
9. Luhn, "Warming Climate and Arctic Gas Push."
10. Staalesen, "In Russia Tundra Tragedy." Of course, Gazprom refutes this claim. Most likely the layer of ice over the lichen was a result of a warm and rainy winter caused by global warming.

11. Nikita Zimov, e-mail communication with author, March 14, 2021.
12. At the time of this writing, in June 2020, the summer season is well under way, although this year there are no scientists, no journalists, and certainly no human geographers. Covid-19 has killed thousands, closed borders, and confined millions to their homes. No one has been allowed to visit NESS this year, and although the Zimov family has managed to get to Chersky, losing an entire season of funding will surely hit the project hard. The motif of Chersky as an island becomes starker at a time of planetary immobility in a pandemic. Scientists cannot get in, and Chersky residents cannot get out. I do not know whether people have died there from Covid-19, but I do know that this fragile Arctic community will be rocked by any instability, given the decimation of public funds and crumbling infrastructure. Chersky has no hospital, no real roads, and no way in or out if the planes stop flying.
13. Steffen et al., "The Anthropocene."
14. Colossal, "De-extinction"; Revive and Restore, "News."
15. Brand, "We Are as Gods."
16. Beer et al., "Protection of Permafrost Soils"; Popov, "The Current State of the Pleistocene Park."
17. Van Dooren, *Flight Ways,* 12.
18. Chrulew, "The Mamontogist's Tale." I would like to thank Matthew Chrulew for providing me with a copy of his story in March 2021.
19. Salih and Corry, "Displacing the Anthropocene."
20. Hartman, "The End of White Supremacy."
21. Roth, "Russian Mining Firm Accused."
22. Parreñas, *Decolonizing Extinction,* 9.

Bibliography

Adams, William. "Geographies of Conservation I: De-extinction and Precision Conservation." *Progress in Human Geography* 41, no. 4 (2017): 534–45.

Adams, William, and Shane McCorristine. "Ghost Species: Spectral Geographies of Biodiversity Conservation." *Cultural Geographies* 27, no. 1 (2020): 101–15.

Alex, Bridget. "Something Stirs: What Happens When Permafrost Thaws?" *Discover Magazine,* June 2018. http://discovermagazine.com.

Alexander, Heather, Sue Natali, Mike Loranty, Sarah Ludwig, Valentin Spektor, Sergei Davydov, Nikita Zimov, Ivonne Trujillo, and Michelle C. Mack. "Impacts of Increased Soil Burn Severity on Larch Forest Regeneration on Permafrost Soils of Far Northeastern Siberia." *Forest Ecology and Management* 417 (2018): 144–53.

Aliyev, Nurlan. "Development in Difficult Times: Russia's Arctic Policy through 2035." *Russian Analytical Digest,* no. 256 (September 5, 2020): 2–6. https://doi.org/10.3929/ethz-b-000440622.

Andersen, Ross. "Welcome to Pleistocene Park." *The Atlantic,* April 2017. https://www.theatlantic.com.

Anderson, David. "Primal Life Force under the Ice." *Times Higher Education,* August 8, 2003. https://www.timeshighereducation.com.

Antonova, Maria. "Siberian Region Fights to Preserve Permafrost as Planet Warms." Phys.org, December 5, 2018. https://phys.org.

Aradau, Claudia, and Rens van Munster. *Politics of Catastrophe: Genealogies of the Unknown.* London: Routledge, 2009.

Arkhangelskaya, Svetlana. "The Siberian Plague." *Russia beyond the Headlines,* August 2016. https://www.rbth.com.

Avery, Mark. *A Message from Martha: The Extinction of the Passenger Pigeon and Its Relevance Today.* London: Bloomsbury, 2014.

Bailes, Kendall. "Soviet Science in the Stalin Period: The Case of V. I. Vernadsky and His Scientific School, 1928–1945." *Slavic Review* 45, no. 1 (Spring 1986): 20–37.

Bakermans, Corien, Alexandre I. Tsapin, Virginia Souza-Egipsy, David A. Gilichinsky, and Kenneth H. Nealson. "Reproduction and Metabolism at –10°C of Bacteria Isolated from Siberian Permafrost." *Environmental Microbiology* 5, no. 4 (April 2003): 321–26.

Balmford, Andrew. *Wild Hope: On the Front Lines of Conservation Success.* Chicago: University of Chicago Press, 2012.

Barad, Karen. *Meeting the Universe Halfway: Quantum Physics and the Entanglement of Matter and Meaning.* Durham, N.C.: Duke University Press, 2007.

Barad, Karen. "Posthumanist Performativity: Toward an Understanding of How Matter Comes to Matter." *Signs* 28, no. 3 (2003): 801–31.

Bastian, Michelle. "Fatally Confused: Telling the Time in the Midst of Ecological Crises." *Environmental Philosophy* 9, no. 1 (2012): 23–48.

Bastian, Michelle. "Time." In *Migration: A COMPAS Anthology,* edited by Bridget Anderson and Michael Keith, 52–53. Oxford: Centre on Migration, Policy and Society, 2014.

Bear, Christopher, and Lewis Holloway. "Beyond Resistance: Geographies of Divergent More-Than-Human Conduct in Robotic Milking." *Geoforum* 104 (2019): 212–21.

Beckett, Jan, and Joseph Singer. *Pana O'ahu: Sacred Stones, Sacred Land.* Honolulu: University of Hawai'i Press, 1999.

Bekoff, Marc. *Rewilding Our Hearts: Building Pathways of Compassion and Coexistence.* San Francisco: New World Library, 2014.

Beer, Christian, Nikita Zimov, Johan Olofsson, Philipp Porada, and Sergey Zimov. "Protection of Permafrost Soils from Thawing by Increasing Herbivore Density." *Scientific Reports* 10 (March 2020). https://doi.org/10.1038/s41598-020-60938-y.

Bennett, Elene M., Martin Solan, Reinette Biggs, Timon McPhearson, Albert Norstrom, Per Olsson, Laura Pereira, et al. "Bright Spots: Seeds of a Good Anthropocene." *Frontiers in Ecology and the Environment* 14, no. 8 (October 2016): 441–48.

Bennett, Jane. *Vibrant Matter: A Political Ecology of Things.* Durham, N.C.: Duke University Press, 2009.

Bennett, Mia. "Putin Opens Bridge to Crimea, Leaving a Siberian City Hanging." Cryopolitics, May 16, 2018. https://www.cryopolitics.com.

Bernstein, Anya. *The Future of Immortality: Remaking Life and Death in Contemporary Russia.* Princeton, N.J.: Princeton University Press, 2019.

Blinnikov, Mikhail S., Benjamin V. Gaglioti, Donald A. Walker, Matthew J. Wooller, and Grant D. Zazula. 2011. "Pleistocene Graminoid-Dominated Ecosystems in the Arctic." *Quaternary Science Reviews* 30, nos. 21–22 (2011): 2906–29.

Bollinger, Martin J. *Stalin's Slave Ships: Kolyma, the Gulag Fleet, and the Role of the West.* Westport, Conn.: Greenwood, 2003.

Bolotova, Alla. "Colonization of Nature in the Soviet Union: State Ideology, Public Discourse, and the Experience of Geologists." *Historical Social Research* 29, no. 3 (2004): 104–23.

Brand, Stewart. "We Are as Gods and Have to Get Good at It: Stewart Brand

Talks about His Ecopragmatist Manifesto." Interview by John Brockman. Edge, August 18, 2009. Video, 20:11. https://www.edge.org.
Brand, Stewart. *Whole Earth Discipline: Why Dense Cities, Nuclear Power, Transgenic Crops, Restored Wildlands, and Geoengineering Are Necessary.* New York: Penguin, 2009.
Braun, Bruce. "Biopolitics and the Molecularization of Life." *Cultural Geographies* 14, no. 1 (2007): 6–28.
Braun, Bruce, and Sarah J. Whatmore. "The Stuff of Politics: An Introduction." In *Political Matter: Technoscience, Democracy, and Public Life,* edited by Bruce Braun and Sarah J. Whatmore, ix–xxxix. Minneapolis: University of Minnesota Press, 2010.
Braverman, Irus. "En-listing Life: Red Is the Color of Threatened Species Lists." In *Critical Animal Geographies: Politics, Intersections, and Hierarchies in a Multispecies World,* edited by Kathryn Gillespie and Rosemary-Clair Collard, 184–202. London: Routledge, 2015.
Bravo, Michael. "A Cryopolitics to Reclaim Our Frozen Material States." In Radin and Kowal, *Cryopolitics,* 27–57.
Breithoff, Esther, and Rodney Harrison. "From Ark to Bank: Extinction, Proxies and Biocapitals in Ex-Situ Biodiversity Conservation Practices." *International Journal of Heritage Studies* 26, no. 1 (2020): 37–55.
Brook, Barry W., and David M. J. S. Bowman. "The Uncertain Blitzkrieg of Pleistocene Megafauna." *Journal of Biogeography* 31, no. 4 (April 2004): 517–23.
Bruno, Andy. *The Nature of Soviet Power: An Arctic Environmental History.* Cambridge: Cambridge University Press, 2016.
Bruyere, Vincent. "Puss-in-Boots Goes to Pleistocene Park." *Oxford Literary Review* 41, no. 1 (2019): 127–40.
Cameron, Emilie. *Far Off Metal River: Inuit Lands, Settler Stories, and the Making of the Contemporary Arctic.* Vancouver: UBC Press, 2015.
Carey, Mark. "The History of Ice: How Glaciers Became an Endangered Species." *Environmental History* 12, no. 3 (July 2007): 497–527.
Carey, Mark, M. Jackson, Alessandro Antonello, and Jaclyn Rushing. "Glaciers, Gender, and Science: A Feminist Glaciology Framework for Global Environmental Change Research." *Progress in Human Geography* 40, no. 6 (2016): 770–93.
Carsten, Janet. "Introduction: Blood Will Out." *Journal of the Royal Anthropological Institute* 19, no. S1 (May 2013): 1–23.
Ceballos, Gerado, Paul R. Ehrlich, Anthony D. Barnosky, Andrés Garcia, Robert M. Pringle, and Todd M. Palmer. "Accelerated Modern Human-Induced Species Loss: Entering the Sixth Mass Extinction." *Science Advances* 1, no. 5 (June 2015): 1–5.
Chakrabarty, Dipesh. "Anthropocene Time." *History and Theory* 57, no. 1 (March 2018): 5–32.
Chakrabarty, Dipesh. "The Climate of History: Four Theses." *Critical Inquiry* 35, no. 2 (2009): 197–222.

Chakrabarty, Dipesh. "Critical Zones." Paper presented at the virtual conference "Observatories for Earthly Politics," May 22–24, 2020.

Chapple, Amos. "The Mammoth Pirates." Radio Free Europe, August 23, 2016. https://www.rferl.org.

Chrulew, Matthew. "Freezing the Ark: The Cryopolitics of Endangered Species Preservation." In Radin and Kowal, *Cryopolitics*, 283–305.

Chrulew, Matthew. "The Mamontogist's Tale." *Cosmos*, December 2015. https://cosmosmagazine.com (no longer available online).

Chu, Pey-Yi. *The Life of Permafrost: A History of Frozen Earth in Russian and Soviet Science*. Toronto: University of Toronto Press, 2021.

Chu, Pey-Yi. "Mapping Permafrost Country: Creating an Environmental Object in the Soviet Union, 1920s–1940s." *Environmental History* 20, no. 3 (July 2015): 396–421.

Church, George, and Ed Regis. *Regenesis: How Synthetic Biology Will Reinvent Nature*. New York: Basic Books, 2012.

Clark, Nigel. *Inhuman Nature: Sociable Life on a Dynamic Planet*. London: Sage, 2010.

Clark, Nigel. "Politics of Strata." *Theory, Culture & Society* 34, nos. 2–3 (2017): 211–31.

Cohen, Claudine. *The Fate of the Mammoth: Fossils, Myth, and History*. Chicago: University of Chicago Press, 2002.

Cohen, Jeffrey Jerome. *Stone: An Ecology of the Inhuman*. Minneapolis: University of Minnesota Press, 2015.

Cohen, Shlomo. "The Ethics of De-extinction." *Nanoethics* 8 (2014): 165–78.

Colebrook, Claire. *Death of the Posthuman*. Ann Arbor, Mich.: Open Humanities Press, 2014.

Colebrook, Claire. "The Future in the Anthropocene: Extinction and the Imagination." In *Climate and Literature*, edited by Adeline Johns-Putra, 263–80. Cambridge: Cambridge University Press, 2019.

Colebrook, Claire. "The Time of Planetary Memory." *Textual Practice* 3, no. 5 (2017): 1017–24.

Collard, Rosemary-Claire, and Jessica Dempsey. "Capitalist Natures in Five Orientations." *Capitalism Nature Socialism* 28, no. 1 (2017): 78–97.

Collins, Patricia Hill. "Learning from the Outsider Within: The Sociological Significance of Black Feminist Thought." *Social Problems* 33, no. 6 (Winter 1986): 14–32.

Colossal. "De-extinction." Accessed May 27, 2022. https://colossal.com.

Colossal. "The Mammoth." Accessed May 27, 2022. https://colossal.com.

Coole, Diana, and Samantha Frost. "Introducing the New Materialisms." In *New Materialisms: Ontology, Agency, and Politics*, edited by Diana Coole and Samantha Frost, 1–43. Durham, N.C.: Duke University Press, 2010.

Cormier, Zoe. "Why the Arctic Is Smouldering." BBC Future, August 26, 2019. https://www.bbc.com/future/article/20190822-why-is-the-arctic-on-fire.

Corradi, C., Olaf Kolle, K. Walter, Sergey Zimov, and Ernst D. Schulze. "Carbon Dioxide and Methane Exchange of a North-East Siberian Tussock Tundra." *Global Change Biology* 11, no. 11 (November 2005): 1910–25.

Cosgrove, Denis, and Veronica Della Dora. "Introduction: High Places." In *High Places: Cultural Geographies of Mountains, Ice and Science,* edited by Denis Cosgrove and Veronica Della Dora, 1–19. London: I. B. Tauris, 2009.

Crate, Susan, Matthias Ulrich, J. Otto Habeck, Aleksey R. Desyatkin, Roman V. Desyatkin, Alexander N. Fedorov, Tetsuya Hiyama, et al. "Permafrost Livelihoods: A Transdisciplinary Review and Analysis of Thermokarst-Based Systems of Indigenous Land Use." *Anthropocene* 18 (June 2017): 89–104.

Crist, Eileen. "On the Poverty of Our Nomenclature." *Environmental Humanities* 3, no. 1 (2013): 129–47.

Cruikshank, Julie. *Do Glaciers Listen? Local Knowledge, Colonial Encounters, and Social Imagination.* Vancouver: UBC Press, 2005.

Crutzen, Paul J., and Eugene F. Stoermer. "The Anthropocene." *Global Change Newsletter* 41 (2000): 17–18.

Cunha, Daniel. "The Geology of the Ruling Class?" *Anthropocene Review* 2, no. 3 (2015): 262–66.

Danowski, Déborah, and Eduardo Viveiros de Castro. *The Ends of the World.* Translated by Rodrigo Nunes. Cambridge: Polity Press, 2016.

Davis, Heather, and Zoe Todd. "On the Importance of a Date, or, Decolonizing the Anthropocene." *Acme* 16, no. 4 (2017): 761–80.

Dawson, Ashley. "Biocapitalism and De-extinction." In Grusin, *After Extinction,* 173–200.

de Goede, Marieke, and Samuel Randalls. "Precaution, Preemption: Arts and Technologies of the Actionable Future." *Environment and Planning D: Society and Space* 27, no. 5 (2009): 859–78.

de la Cadena, Marisol, and Mario Blaser. "Pluriverse." In de la Cadena and Blaser, *A World of Many Worlds,* 1–23.

de la Cadena, Marisol, and Mario Blaser, eds. *A World of Many Worlds.* Durham, N.C.: Duke University Press, 2018.

DeLoughrey, Elizabeth M. *Allegories of the Anthropocene.* Durham, N.C.: Duke University Press, 2019.

Deniau, Jean-Charles, dir. *Raising the Mammoth.* Silver Spring, Md.: Discovery Channel/France 3/NOVI Productions, 2000.

Dennis, Michael Aaron. "A Polar Perspective." In *Globalizing Polar Science: Reconsidering the International Polar and Geophysical Years,* edited by Roger D. Launius, James R. Fleming, and David H. DeVorkin, 13–23. New York: Palgrave Macmillan, 2010.

DeSilvey, Caitlin. *Curated Decay: Heritage beyond Saving.* Minneapolis: University of Minnesota Press, 2017.

DeSilvey, Caitlin, and Tim Edensor. "Reckoning with Ruins." *Progress in Human Geography* 37, no. 4 (2013): 465–85.

Devlin, Hannah. "Woolly Mammoth on Verge of Resurrection, Scientists Reveal." *The Guardian,* February 16, 2017. https://www.theguardian.com.

Dobinski, Wojciech. "Permafrost." *Earth Science Reviews* 108, nos. 3–4 (October 2011): 158–69.

Dodds, Klaus. "Geopolitics and Ice Humanities: Elemental, Metaphorical and Volumetric Reverberations." *Geopolitics* 26, no. 4 (2021): 1121–49.

Doel, Ronald E. "Constituting the Cold War Earth Sciences: The Military's Influence on the Earth Sciences in the USA after 1945." *Social Studies of Science* 33, no. 5 (2003): 635–66.

Drake, Nadia. "Elon Musk: A Million Humans Could Live on Mars by the 2060s." *National Geographic,* September 28, 2016. https://www.nationalgeographic.com.

Dutta, Koshik, E. A. Schuur, J. C. Neff, and Sergey Zimov. "Potential Carbon Release from Permafrost Soils of Northeastern Siberia." *Global Change Biology* 12, no. 12 (December 2006): 2336–51.

Edelman, Lee. *No Future: Queer Theory and the Death Drive.* Durham, N.C.: Duke University Press, 2004.

Elden, Stuart. "Land, Terrain, Territory." *Progress in Human Geography* 34, no. 6 (2010): 799–817.

Elden, Stuart. "Terrain, Politics, History." *Dialogues in Human Geography* 11, no. 2 (July 2021): 170–89.

EXO-YKT. "The Ice Has Broken: The Point of No Return." August 28, 2017. https://www.exo-ykt.ru/articles/02/612/19438.

Farman, Abou. "Cryonic Suspension as Eschatological Technology in the Secular Age." In *A Companion to the Anthropology of Death,* edited by Antonius C. G. M. Robben, 307–19. Oxford: John Wiley, 2018.

Farman, Abou. *On Not Dying: Secular Immortality in the Age of Technoscience.* Minneapolis: University of Minnesota Press, 2020.

Farnham, Timothy. *Saving Nature's Legacy: Origins of the Idea of Biological Diversity.* New Haven, Conn.: Yale University Press, 2007.

Farrier, David. *Anthropocene Poetics: Deep Time, Sacrifice Zones, and Extinction.* Minneapolis: University of Minnesota Press, 2018.

Farrier, David. *Footprints: In Search of Future Fossils.* New York: Macmillan, 2020.

Federici, Silvia. *Re-enchanting the World: Feminism and the Politics of the Commons.* Oakland, Calif.: PM Press, 2019.

Fedorov, Alexander N., Nikolay F. Vasilyev, Yaroslav I. Torgovkin, Alena A. Shestakova, Stepan P. Varlamov, Mikhail N. Zheleznyak, Viktor V. Shepelev, et al. "Permafrost-Landscape Map of the Republic of Sakha (Yakutia) on a Scale 1:1,500,000." *Geosciences* 8, no. 12 (December 2018). https://doi.org/10.3390/geosciences8120465.

Feinberg, Matthew, and Rob Willer. "Apocalypse Soon? Dire Messages Reduce Belief in Global Warming by Contradicting Just-World Beliefs." *Psychological Science* 22, no. 1 (2011): 34–38.

Fletcher, Amy Lynn. *De-extinction and the Genomics Revolution: Life on Demand.* Cham, Switzerland: Palgrave Macmillan, 2020.

Forsyth, James. *A History of the Peoples of Siberia: Russia's North Asian Colony, 1581–1990.* Cambridge: Cambridge University Press, 1992.

Foucault, Michel. *The History of Sexuality.* Vol. 1, *An Introduction.* New York: Pantheon, 1980.

Franklin, Sarah. "From Blood to Genes: Rethinking Consanguinity in the Context of Geneticization." In *Blood and Kinship: Matter for Metaphor from Ancient Rome to the Present,* edited by Christopher H. Johnson, Bernhard Jussen, David Warren Sabean, and Simon Teuscher, 285–306. Oxford: Berghahn Books, 2013.

Franklin, Sarah. "Life Itself: Global Nature and the Genetic Imaginary." In *Global Nature, Global Culture,* edited by Sarah Franklin, Celia Lury, and Jackie Stacey, 188–227. London: Sage, 2000.

Frei, Christian, dir. *Genesis 2.0.* Bern: Rise and Shine Films, 2018.

Froese, Duane G., John A. Westgate, Alberto V. Reyes, Randolph J. Enkin, and Shari J. Preece. "Ancient Permafrost and a Future, Warmer Arctic." *Science* 321, no. 5896 (2008): 1648.

Frozen Ark. "Future Proofing." Accessed January 12, 2018. https://frozenark.org.

Garlick, Ben. "Cultural Geographies of Extinction: Animal Culture among Scottish Ospreys." *Transactions of the Institute of British Geographers* 44, no. 2 (2019): 226–41.

Garlick, Ben, and Kate Symons. "Geographies of Extinction: Exploring the Spatiotemporal Relations of Species Death." *Environmental Humanities* 12, no. 1 (2020): 296–320.

Geertz, Clifford. *The Interpretation of Cultures: Selected Essays.* New York: Basic Books, 1973.

Geissler, P. Wenzel, and Ann H. Kelly. "A Home for Science: The Life and Times of Tropical and Polar Field Stations." *Social Studies of Science* 46, no. 6 (2016): 797–808.

Gertcyk, Olga. "Huge Cull of 250,000 Reindeer by Christmas in Yamalo-Nenets after Anthrax Outbreak." *Siberian Times,* September 19, 2016. https://siberiantimes.com.

Giffney, Noreen, and Myra Hird. "Introduction: Queering the Non/Human." In *Queering the Non/human,* edited by Noreen Giffney and Myra Hird, 1–16. Aldershot: Ashgate, 2008.

Ginn, Franklin. "When Horses Won't Eat: Apocalypse and the Anthropocene." *Annals of the Association of American Geographers* 105, no. 2 (2015): 351–59.

Gordillo, Gastón. "The Power of Terrain: The Affective Materiality of Planet Earth in the Age of Revolution." *Dialogues in Human Geography* 11, no. 2 (July 2021): 190–94.

Gordillo, Gastón. "The Void: Invisible Ruins on the Edges of Empire." In Stoler, *Imperial Debris,* 227–51.

Gordon, Avery F. *Ghostly Matters: Haunting and the Sociological Imagination.* Minneapolis: University of Minnesota Press, 1997.
Grant, Bruce. *In the Soviet House of Culture: A Century of Perestroikas.* Princeton, N.J.: Princeton University Press, 1996.
Griggs, Jessica. "Gene Editing: Bring It On." *New Scientist,* September 18, 2015. https://www.newscientist.com.
Grosz, Elizabeth. *Becoming Undone: Darwinian Reflections on Life, Politics, and Art.* Durham, N.C.: Duke University Press, 2011.
Grosz, Elizabeth, Kathryn Yusoff, and Nigel Clark. "An Interview with Elizabeth Grosz: Geopower, Inhumanism and the Biopolitical." *Theory, Culture & Society* 34, nos. 2–3 (2017): 129–46.
Grove, Jairus. "Of an Apocalyptic Tone Recently Adopted in Everything: The Anthropocene or Peak Humanity?" *Theory & Event* 18, no. 3 (2015). https://www.muse.jhu.edu.
Grusin, Richard, ed. *After Extinction.* Minneapolis: University of Minnesota Press, 2018.
Guillame, Bertrand. "Vernadsky's Philosophical Legacy: A Perspective from the Anthropocene." *Anthropocene Review* 1, no. 2 (2014): 137–46.
Hamilton, Clive. *Earthmasters: The Dawn of the Age of Climate Engineering.* New Haven, Conn.: Yale University Press, 2013.
Haraway, Donna. "Anthropocene, Capitalocene, Plantationocene, Chthulucene: Making Kin." *Environmental Humanities* 6, no. 1 (2015): 159–65.
Haraway, Donna. "Situated Knowledges: The Science Question in Feminism and the Privilege of Partial Perspective." *Feminist Studies* 14, no. 3 (1988): 575–99.
Haraway, Donna. *Staying with the Trouble: Making Kin in the Chthulucene.* Durham, N.C.: Duke University Press, 2016.
Haraway, Donna. *When Species Meet.* Minneapolis: University of Minnesota Press, 2008.
Haraway, Donna, Noboru Ishikawa, Scott F. Gilbert, Kenneth Olwig, Anna L. Tsing, and Nils Bubandt. "Anthropologists Are Talking—about the Anthropocene." *Ethnos* 81, no. 3 (2015): 535–64.
Harries, John. "Of Bleeding Skulls and the Postcolonial Uncanny: Bones and the Presence of Nonosabasut and Demasduit." *Journal of Material Culture* 15, no. 4 (2010): 403–21.
Hartman, Saidiya. "The End of White Supremacy, an American Romance." *BOMB Magazine,* June 5, 2020. https://bombmagazine.org.
Hastrup, Kirsten, and Martin Skrydstrup, eds. *The Social Life of Climate Change Models: Anticipating Nature.* New York: Routledge, 2013.
Heise, Ursula. *Imagining Extinction: The Cultural Meanings of Endangered Species.* Chicago: University of Chicago Press, 2017.
Heise, Ursula. *Sense of Place and Sense of Planet: The Environmental Imagination of the Global.* Oxford: Oxford University Press, 2008.
Heleniak, Timothy, Eeva Turunen, and Shinan Wang. "Cities on Ice:

Population Change in the Arctic." Northern Forum, October 26, 2018. https://www.northernforum.org.

Hird, Myra. "Knowing Waste: Towards an Inhuman Epistemology." *Social Epistemology* 23, nos. 3–4 (2012): 453–69.

Hoag, Hannah. "Permafrost That Lives Up to Its Name." *Nature*, September 18, 2008. https://doi.org/10.1038/news.2008.1119.

Holbraad, Martin, Bruce Kapferer, and Julia F. Sauma. "Introduction: Critical Ruptures." In *Ruptures: Anthropologies of Discontinuity in Times of Turmoil*, edited by Martin Holbraad, Bruce Kapferer, and Julia F. Sauma, 1–26. London: UCL Press, 2019.

Iida, Mayu. "Fukushima: A Tale about Beginnings and Ends, or How the Disaster Has Become Truly Environmental." *Feminist Review* 118 (2018): 93–99.

Ingold, Tim. *Hunters, Pastoralists and Ranchers: Reindeer Economies and Their Transformations.* Cambridge: Cambridge University Press, 1980.

Ingold, Tim. "A Northern Ontology?" Paper presented at the conference "History of Arctic Anthropology," Royal Anthropological Institute, London, February 27, 2020.

Ingold, Tim. *The Perception of the Environment: Essays on Livelihood, Dwelling and Skill.* London: Routledge, 2000.

Intelligence Squared U.S. "Don't Bring Extinct Creatures Back to Life." January 31, 2019. https://www.intelligencesquaredus.org.

Intergovernmental Panel on Climate Change. *Global Warming of 1.5°C: An IPCC Special Report on the Impacts of Global Warming of 1.5°C above Pre-industrial Levels and Related Global Greenhouse Gas Emission Pathways, in the Context of Strengthening the Global Response to the Threat of Climate Change, Sustainable Development, and Efforts to Eradicate Poverty.* Geneva: World Meteorological Organization, 2018.

Intergovernmental Panel on Climate Change. *Special Report on the Ocean and Cryosphere in a Changing Climate.* Geneva: World Meteorological Organization, 2019.

International Permafrost Association. "What Is Permafrost?" Accessed January 10, 2022. https://www.permafrost.org.

International Union for Conservation of Nature. "IUCN Red List of Threatened Species." Accessed January 10, 2022. https://www.iucnredlist.org.

Jackson, M. "Glaciers and Climate Change: Narratives of Ruined Futures." *WIREs Climate Change* 6, no. 5 (2015): 479–92.

Jiménez, Alberto Corsín, and Rane Willerslev. "'An Anthropological Concept of the Concept': Reversibility among the Siberian Yukaghirs." *Journal of the Royal Anthropological Institute* 13, no. 3 (September 2007): 527–44.

Jørgensen, Dolly. "Endling, the Power of the Last in an Extinction-Prone World." *Environmental Philosophy* 14, no. 1 (2017): 119–38.

Jørgensen, Dolly. *Recovering Lost Species in the Modern Age: Histories of Longing and Belonging.* Cambridge: MIT Press, 2019.

Jørgensen, Dolly. "Rethinking Rewilding." *Geoforum* 65 (2015): 482–88.

Josephson, Paul R. *The Conquest of the Russian Arctic.* Cambridge: Harvard University Press, 2014.

Kearnes, Matthew, and Lauren Rickards. "Earthly Graves for Environmental Futures: Techno-burial Practices." *Futures* 92 (September 2017): 48–58.

Kelly, Casey Ryan. "The Man-pocalypse: *Doomsday Preppers* and the Rituals of Apocalyptic Manhood." *Text and Performance Quarterly* 36, nos. 2–3 (2016): 95–114.

Kintisch, Eli. "These Ice Cellars Fed People for Generations. Now They're Melting." *National Geographic,* October 30, 2015. https://www.nationalgeographic.com.

Kirsch, Scott, and Don Mitchell. "The Nature of Things: Dead Labor, Nonhuman Actors, and the Persistence of Marxism." *Antipode* 36, no. 4 (September 2004): 687–705.

Krmpotich, Cara, Joost Fontein, and John Harries. "The Substance of Bones: The Emotive Materiality and Affective Presence of Human Remains." *Journal of Material Culture* 15, no. 4 (2010): 371–84.

Krupnik, Igor, and Dyanna Jolly. "Introduction." In *The Earth Is Faster Now: Indigenous Observations of Arctic Environment Change,* edited by Igor Krupnik and Dyanna Jolly, 1–11. Fairbanks: Arctic Research Consortium of the United States/Smithsonian Institution Arctic Studies Center, 2002.

Lachenbruch, Arthur H., and B. Vaughn Marshall. "Changing Climate: Geothermal Evidence from Permafrost in the Alaskan Arctic." *Science* 234, no. 4777 (1986): 689–96.

Landecker, Hannah. *Culturing Life: How Cells Became Technologies.* Cambridge, Mass.: Harvard University Press, 2007.

Lara, Mark J., Hélène Genet, Anthony D. McGuire, Eugénie S. Euskirchen, Yujin Zhang, Dana R. N. Brown, Mark T. Jorgenson, Vladimir Romanovsky, Amy Breen, and William R. Bolton. "Thermokarst Rates Intensify Due to Climate Change and Forest Fragmentation in an Alaskan Boreal Forest Lowland." *Global Change Biology* 22, no. 2 (February 2016): 816–29.

Laruelle, Marlène. "Russia's Arctic Policy: A Power Strategy and Its Limits." *Russie.Nei.Visions,* no. 117 (March 2020). https://www.ifri.org.

Leslie, Esther. *Liquid Crystals: The Science and Art of a Fluid Form.* London: Reaktion Books, 2018.

Levy, Sharon. *Once and Future Giants: What Ice Age Extinctions Tell Us about the Fate of Earth's Largest Animals.* Oxford: Oxford University Press, 2011.

Liu, Miaojia, and Jacob Kronbak. "The Potential Economic Viability of Using the Northern Sea Route (NSR) as an Alternative Route between

Asia and Europe." *Journal of Transport Geography* 18, no. 3 (May 2010): 434–44.
Livingstone, David. *Putting Science in Its Place: Geographies of Scientific Knowledge.* Chicago: University of Chicago Press, 2003.
Lohr, Barbara, dir. *Siberia: The Melting Permafrost.* Strasbourg: Arte TV, 2017.
Loranty, Michael M., Logan T. Berner, Eric D. Taber, Heather Kropp, Susan M. Natali, Heather D. Alexander, Sergei P. Davydov, and Nikita S. Zimov. "Understory Vegetation Mediates Permafrost Active Layer Dynamics and Carbon Dioxide Fluxes in Open-Canopy Larch Forests of Northeastern Siberia. *PLOS One* 13, no. 3 (March 2018): 1–17.
Lorimer, Jamie. "Nonhuman Charisma." *Environment and Planning D: Society and Space* 25, no. 5 (2007): 911–32.
Lorimer, Jamie, and Clemens Driessen. "Bovine Biopolitics and the Promise of Monsters in the Rewilding of Heck Cattle." *Geoforum* 48 (2013): 249–59.
Lorimer, Jamie, and Clemens Driessen. "From 'Nazi Cows' to Cosmopolitan 'Ecological Engineers': Specifying Rewilding through a History of Heck Cattle." *Annals of the American Association of Geographers* 106, no. 3 (2016): 631–52.
Lorimer, Jamie, and Clemens Driessen. "Wild Experiments at the Oostvaardersplassen: Rethinking Environmentalism in the Anthropocene." *Transactions of the Institute of British Geographers* 39, no. 2 (2014): 169–81.
Luhn, Alec. "Anthrax Outbreak Triggered by Climate Change Kills Boy in Arctic Circle." *The Guardian,* August 1, 2016. https://www.theguardian.com.
Luhn, Alec. "Warming Climate and Arctic Gas Push Threaten Russia's Reindeer Herders." Reuters, April 24, 2020. https://www.reuters.com.
Malhi, Yadvinder, Christopher E. Doughty, Mauro Galetti, Felisa A. Smith, Jens-Christian Svenning, and John W. Terborgh. "Megafauna and Ecosystem Function from the Pleistocene to the Anthropocene." *PNAS* 113, no. 4 (2016): 838–46.
Malm, Andreas, and Alf Hornborg. "The Geology of Mankind? A Critique of the Anthropocene Narrative." *Anthropocene Review* 1, no. 1 (2014): 62–69.
McCannon, John. *A History of the Arctic: Nature, Exploration, and Exploitation.* London: Reaktion Books, 2012.
McCannon, John. *Red Arctic: Polar Exploration and the Myth of the North in the Soviet Union, 1932–1939.* Oxford: Oxford University Press, 1998.
McKay, John J. *Discovering the Mammoth: A Tale of Giants, Unicorns, Ivory, and the Birth of a New Science.* New York: Pegasus Books, 2017.
Melnikov Permafrost Institute, Yakutsk, Siberian Branch, Russian Academy of Sciences. "Permafrost Seed Repository." Accessed October 15, 2020. https://mpi.ysn.ru/en.

Melo Zurita, Maria de Lourdes, Paul George Munro, and Donna Houston. "Un-earthing the Subterranean Anthropocene." *Area* 50, no. 3 (2018): 298–305.

Mendes, Margarida. "Molecular Colonialism." In *Matter Fictions,* edited by Margarida Mendes, 127–40. Berlin: Sternberg Press, 2017.

Methmann, Chris, and Delf Rothe. "Politics for the Day after Tomorrow: The Logic of Apocalypse in Global Climate Politics." *Security Dialogue* 43, no. 4 (2012): 323–44.

Meyer, Robinson. "The Zombie Diseases of Climate Change." *The Atlantic,* November 6, 2017. https://www.theatlantic.com.

Milkoreit, Manjana. "Imaginary Politics: Climate Change and the Making of the Future." *Elementa: The Science of the Anthropocene* 5 (2017): 62–75.

Mitchell, Audra. "Decolonizing against Extinction, Part III: White Tears and Mourning." Worldly, December 14, 2017. https://worldlyir .wordpress.com.

Mitchell, Ronald B., William C. Clark, David W. Cash, and Nancy M. Dickson. *Global Environmental Assessments: Information and Influence.* Cambridge: MIT Press, 2016.

Monbiot, George. *Feral.* London: Penguin, 2013.

Moore, Amelia. "Anthropocene Anthropology: Reconceptualizing Contemporary Global Change." *Journal of the Royal Anthropological Institute* 22, no. 1 (March 2016): 27–46.

Moore, Jason W. "Anthropocene or Capitalocene? Nature, History, and the Crisis of Capitalism." In *Anthropocene or Capitalocene? Nature, History, and the Crisis of Capitalism,* edited by Jason W. Moore, 1–13. Oakland, Calif.: PM Press, 2016.

Moynihan, Thomas. *X-Risk: How Humanity Discovered Its Own Extinction.* Falmouth: Urbanomic, 2020.

Murton, Julian B., Mary E. Edwards, Anatoly V. Lozhkin, Patricia M. Anderson, Grigoriy N. Savvinov, Nadezhda Bakulina, Olesya V. Bondarenko, et al. "Preliminary Paleoenvironmental Analysis of Permafrost Deposits at Batagaika Megaslump, Yana Uplands, Northeast Siberia." *Quaternary Research* 87, no. 2 (February 2017): 314–30.

Murton, Julian B., Tomasz Goslar, Mary E. Edwards, Mark D. Bateman, Petr P. Danilov, Grigoriy N. Savvinov, Stanislav V. Gubin, et al. "Palaeoenvironmental Interpretation of Yedoma Silt (Ice Complex) Deposition as Cold-Climate Loess, Duvanny Yar, Northeast Siberia." *Permafrost and Periglacial Processes* 26, no. 3 (2015): 208–88.

Myers, Steven Lee. "Weaning Itself from Elephant Ivory, China Turns to Mammoths." *New York Times,* August 6, 2017. https://www.nytimes .com.

Nakano, Tomoko, Shunich Kuniyoshi, and Masami Fukuda. "Temporal Variation in Methane Emission from Tundra Wetlands in a Permafrost Area, Northeastern Siberia." *Atmospheric Environment* 34, no. 8 (2000): 1205–13.

Nash, J. Madeleine. "The Impermanence of Permafrost." *Hakai Magazine*, March 27, 2018. https://www.hakaimagazine.com.

Natali, Susan M., John P. Holdren, Brendan M. Rogers, Rachael Treharne, Philip B. Duffy, Rafe Pomerance, and Erin Macdonald. "Permafrost Carbon Feedbacks Threaten Global Climate Goals." *PNAS* 118, no. 21 (2021). https://doi.org/10.1073/pnas.2100163118.

National Research Council of Canada. *Glossary of Permafrost and Related Ground-Ice Terms*. Ottawa: National Research Council of Canada, 1988.

Neslen, Arthur. "US Military Agency Invests $100m in Genetic Extinction Technologies." *The Guardian*, December 4, 2017. https://www.theguardian.com.

Nettheim, Daniel, dir. *The Hunter*. East Melbourne: Madman Entertainment, 2011.

Nixon, Rob. "The Anthropocene: The Promise and Pitfalls of an Epochal Idea." *Edge Effects*, November 6, 2014. https://edgeeffects.net.

Northcott, Michael. "Eschatology in the Anthropocene: From the *Chronos* of Deep Time to the *Kairos* of the Age of Humans." In *The Anthropocene and the Global Environmental Crisis: Rethinking Modernity in a New Epoch*, edited by Clive Hamilton, Christophe Bonneuil, and François Gemenne, 100–111. London: Routledge, 2015.

Novak, Ben Jacob. "De-extinction." *Genes* 9, no. 11 (2018). https://doi.org/10.3390/genes9110548.

Obruchev, Vladimir. *Plutonia* [in Russian]. Winnipeg: Raduga, 1924.

Olson, Valerie, and Lisa Messeri. "Beyond the Anthropocene: Un-earthing an Epoch." *Environment and Society: Advances in Research* 6 (2015): 28–47.

O'Reilly, Jessica. "Sensing the Ice: Field Science, Models, and Expert Intimacy with Knowledge." *Journal of the Royal Anthropological Institute* 21, no. S1 (April 2016): 27–45.

Palkopoulou, Eleftheria, Swapan Mallick, Pontus Skoglund, Jacob Enk, Nadin Rohland, Heng Li, Ayça Omrak, et al. "Complete Genomes Reveal Signatures of Demographic and Genetic Declines in the Woolly Mammoth." *Current Biology* 25, no. 10 (May 2015): 1395–1400.

Palmer, Douglas. *Fossil Revolution: The Finds That Changed Our View of the Past*. London: Trafalgar Square, 2003.

Parikka, Jussi. "Planetary Memories: After Extinction, the Imagined Future." In Grusin, *After Extinction*, 27–51.

Parreñas, Juno Salazar. *Decolonizing Extinction: The Work of Care in Orangutan Rehabilitation*. Durham, N.C.: Duke University Press, 2018.

Pauly, Daniel. "Anecdotes and the Shifting Baselines of Fisheries." *Trends in Ecology and Evolution* 10, no. 10 (1995): 430.

PCF Action Group. "Are Permafrost Thaw Interventions Possible?" PCF Dialogue 3. Virtual discussion presented by the Cascade Institute and the Canadian Permafrost Association, March 18, 2021. https://cascadeinstitute.org.

Pearson, Chris. "Beyond 'Resistance': Rethinking Nonhuman Agency for a More-Than-Human World." *European Review of History* 22, no. 5 (2015): 709–25.

" 'Pleistocene Park' Experiment." BBC News, May 15, 2005. http://news.bbc.co.uk.

Pleistocene Park Foundation. "Bison to Save the World." Indiegogo campaign. Accessed January 23, 2019. https://www.indiegogo.com.

Pleistocene Park Foundation. "Pleistocene Park: An Ice Age Ecosystem to Save the World." Kickstarter campaign. Accessed January 23, 2019. https://www.kickstarter.com.

Popov, Igor. "The Current State of the Pleistocene Park, Russia (an Experiment in the Restoration of Megafauna in a Boreal Environment)." *The Holocene* 30, no. 10 (2020): 1471–73.

Povinelli, Elizabeth A. "Do Rocks Listen? The Cultural Politics of Apprehending Australian Aboriginal Labor." *American Anthropologist* 97, no. 3 (1995): 505–18.

Povinelli, Elizabeth A. *The Empire of Love: Toward a Theory of Intimacy, Genealogy, and Carnality.* Durham, N.C.: Duke University Press, 2006.

Povinelli, Elizabeth A. *Geontologies: A Requiem to Late Liberalism.* Durham, N.C.: Duke University Press, 2016.

Povinelli, Elizabeth A., Julieta Aranda, Brian Kuan Wood, and Anton Vidokle. "Quasi-events." *e-flux Journal,* no. 58 (October 2014). https://www.e-flux.com.

Powell, Nick Clarke, dir. *Ice Age: Return of the Mammoth.* London: Channel 4, 2019.

Powell, Nick Clarke, dir. *Woolly Mammoth: The Autopsy.* London: Channel 4, 2014.

Premiyak, Liza. "The Last Frontier: Photographing the Final Days of Winter Sun in the Arctic Town of Chersky." *Calvert Journal,* December 27, 2016. https://www.calvertjournal.com.

Pugh, Jonathan, and David Chandler. *Anthropocene Islands: Entangled Worlds.* London: University of Westminster Press, 2021.

Pushkareva, E. T., and A. A. Burykin. Fol'klor narodov Severa (kul'turno-antropologicheskie aspekty) [Folklore of the peoples of the North (cultural and anthropological aspects)]. St. Petersburg: Petersburg Oriental Studies, 2011.

Putin, Vladimir. "O strategii razvitiya Arkticheskoy zony Rossiyskoy Federatsii i obespecheniya natsional'noy bezopasnosti na period do 2035 goda" [On the strategy for the development of the Arctic Zone of the Russian Federation and ensuring national security for the period until 2035]." Government of the Russian Federation, October 26, 2020. https://www.arctic2035.ru.

Radin, Joanna. *Life on Ice: A History of New Uses for Cold Blood.* Chicago: University of Chicago Press, 2017.

Radin, Joanna, and Emma Kowal, eds. *Cryopolitics: Frozen Life in a Melting World.* Cambridge: MIT Press, 2017.

Radin, Joanna, and Emma Kowal. "Introduction: The Politics of Low Temperature." In Radin and Kowal, *Cryopolitics,* 3–26.

Rajan, Kaushik Sunder. *Biocapital: The Constitution of Postgenomic Life.* Durham, N.C.: Duke University Press, 2006.

Reiter, Berndt. *Constructing the Pluriverse: The Geopolitics of Knowledge.* Durham, N.C.: Duke University Press, 2018.

Republic of Sakha. "Ob okhrane vechnoi merzloty v Respublike Sakha (Yakutiya)" [On the protection of permafrost in the Sakha Republic (Yakutia)], May 22, 2018. https://docs.cntd.ru/document/550111100.

Resnick, Brian. "Melting Permafrost in the Arctic Is Unlocking Diseases and Warping the Landscape." Vox, updated November 15, 2019. https://www.vox.com.

Revive and Restore. "News." Accessed April 30, 2020. https://reviverestore.org.

Rice, Waubgeshig. *Moon of the Crusted Snow.* Toronto: ECW Press, 2018.

Richmond, Douglas J., Mikkel-Holger S. Sinding, and M. Thomas P. Gilbert. "The Potential and Pitfalls of De-extinction." *Zoologica Scripta* 45, no. S1 (October 2016): 22–36.

Rispoli, Giulia, and Jacques Grinevald. "Vladimir Vernadsky and the Co-evolution of the Biosphere, the Noosphere, and the Technosphere." *Technosphere Magazine,* June 20, 2018. https://technosphere-magazine.hkw.de.

Rockström, Johan, Will Steffen, Kevin Noone, Åsa Persson, F. Stuart Chapin III, Eric F. Lambin, Timothy M. Lenton, et al. "A Safe Operating Space for Humanity." *Nature* 461 (2009): 472–75.

Rogacheva, Maria A. "Soviet Scientists Take the Initiative: Khrushchev's Thaw and the Emergence of the Scientific Centre in Chernogolovka." *Europe–Asia Studies* 68, no. 7 (2016): 1179–96.

Romanovsky, Vladimir E., Dmitry Drozdov, Naum G. Oberman, G. V. Malkova, Alexander Khlodov, Sergei Marchenko, N. G. Moskalenko, et al. "Thermal State of Permafrost in Russia." *Permafrost and Periglacial Processes* 21, no. 2 (2010): 136–55.

Rose, Deborah Bird. "In the Shadow of All This Death." In *Animal Death,* edited by Jay Johnston and Fiona Probyn-Rapsey, 1–20. Sydney: Sydney University Press, 2013.

Rose, Deborah Bird. "Multispecies Knots of Ethical Time." *Environmental Philosophy* 9, no. 1 (2012): 127–40.

Rose, Deborah Bird. "Reflections on the Zone of the Incomplete." In Radin and Kowal, *Cryopolitics,* 145–55.

Rose, Nikolas. *The Politics of Life Itself: Biomedicine, Power, and Subjectivity in the Twenty-First Century.* Princeton, N.J.: Princeton University Press, 2007.

Rose, Nikolas. "The Politics of Life Itself." *Theory, Culture & Society* 18, no. 6 (2001): 1–30.

Roth, Andrew. "Russian Mining Firm Accused of Using Global Heating to Avoid Blame for Oil Spill." *The Guardian,* June 9, 2020. https://www.theguardian.com.

Rudwick, Martin J. S. *Bursting the Limits of Time: The Reconstruction of Geohistory in the Age of Revolution.* Chicago: University of Chicago Press, 2005.

Rudwick, Martin J. S. *Earth's Deep History: How It Was Discovered and Why It Matters.* Chicago: University of Chicago Press, 2014.

Russill, Chris, and Zoe Nyssa. "The Tipping Point Trend in Climate Change Communication." *Global Environmental Change* 19, no. 3 (2009): 336–44.

Salazar, Juan Francisco. "Ice Cores as Temporal Probes." *Journal of Contemporary Archaeology* 5, no. 1 (2018): 32–43.

Salazar, Juan Francisco. "Polar Infrastructures." In *The Routledge Companion to Digital Ethnography,* edited by Larissa Hjorth, Heather Horst, Anne Galloway, and Genevieve Bell, 374–83. London: Routledge, 2017.

Salih, Ruba, and Olaf Corry. "Displacing the Anthropocene: Colonisation, Extinction and the Unruliness of Nature in Palestine." *Environment and Planning E: Nature and Space* 5, no. 1 (2022): 381–400.

Sample, Ian. "Oldest *Homo sapiens* Bones Ever Found Shake Foundations of the Human Story." *The Guardian,* June 7, 2017. https://www.theguardian.com.

San Diego Zoo Wildlife Alliance. "White Rhino." Accessed May 30, 2022. https://institute.sandiegozoo.org.

Sandler, Ronald. "The Ethics of Reviving Long Extinct Species." *Conservation Biology* 28, no. 2 (2013): 354–60.

Scanlon, Charles. "Korea's National Shock at Scandal." BBC News, January 13, 2006. http://news.bbc.co.uk.

Schmidt, Jeremy J. "Glacial Deaths, Geologic Extinction." *Environmental Humanities* 13, no. 2 (2021): 281–300.

Schuur, Edward A. G., A. D. McGuire, Christina Schadel, Guido Grosse, J. W. Harden, D. J. Hayes, Gustaf Hugelius, et al. "Climate Change and the Permafrost Carbon Feedback." *Nature* 520 (April 2015): 171–79.

Schweitzer, Peter, Olga Povoroznyuk, and Sigrid Schiesser. "Beyond Wilderness: Towards an Anthropology of Infrastructure and the Built Environment in the Russian North." *Polar Journal* 7, no. 1 (2017): 58–85.

"Scientists 'Confident' That They Can Extract Cells to Clone 42,000 Year Old Extinct Foal." *Siberian Times,* April 8, 2019. https://siberiantimes.com.

Searle, Adam. "Anabiosis and the Liminal Geographies of De/extinction." *Environmental Humanities* 12, no. 1 (2020): 321–45.

Sepkoski, David. *Catastrophic Thinking.* Chicago: University of Chicago Press, 2020.

Sharma, Sarah. *In the Meantime: Temporary and Cultural Politics.* Durham, N.C.: Duke University Press.

Slater, Grant, dir. *Mammoth.* Washington, D.C.: The Atlantic, 2017.

Sleptsov-Sylyk, Nikolai I. *Dykhanie vechnoi merzloty* [The breath of permafrost]. Yakutsk: Bichik, 2013.

Slezkine, Yuri. *Arctic Mirrors: Russia and the Small Peoples of the North.* Ithaca, N.Y.: Cornell University Press, 1994.

Slezkine, Yuri. *The House of Government: A Saga of the Russian Revolution.* Princeton, N.J.: Princeton University Press, 2017.

Smirnov, Nikolay. "The Permafrost." Centre for Experimental Museology, 2016. Accessed June 6, 2022. https://redmuseum.church/en.

Smirnov, Nikolay. "The Provincialization of the Globe and the Trap of Planetarity." *Moscow Art Magazine* 99 (2016): 49–57.

Solly, Meilan. "A Perfectly Preserved 32,000-Year-Old Wolf Head Was Found in Siberian Permafrost." *Smithsonian Magazine,* June 14, 2019. https://www.smithsonianmag.com.

Sörlin, Sverker. "Cryo-history: Ice and the Emerging Arctic Humanities." In *The New Arctic,* edited by Birgitta Evengård, Joan Nymand Larsen, and Øyvind Paasche, 327–39. New York: Springer, 2015.

Spivak, Gayatri Chakravorty. *Death of a Discipline.* New York: Columbia University Press, 2003.

Springgay, Stephanie, and Sarah E. Truman. *Walking Methodologies in a More-Than-Human World.* London: Routledge, 2018.

Squire, Rachel, and Klaus Dodds. "Introduction to the Special Issue: Subterranean Geopolitics." *Geopolitics* 25, no. 1 (2020): 4–16.

Staalesen, Atle. "In Russian Tundra Tragedy, Up to 80,000 Reindeer May Have Starved to Death." *Barents Observer,* March 3, 2021. https://thebarentsobserver.com.

Steffen, Will, Åsa Persson, Lisa Deutsch, Jan Zalasiewicz, Mark Williams, Katherine Richardson, Carole Crumley, et al. "The Anthropocene: From Global Change to Planetary Stewardship." *Ambio* 40, no. 7 (November 2011): 739–761.

Steinberg, Philip, and Kimberley Peters. "Cross-Currents and Undertows: A Response." *Dialogues in Human Geography* 9, no. 3 (November 2019): 333–38.

Steinberg, Philip, and Kimberley Peters. "Wet Ontologies, Fluid Spaces: Giving Depth to Volume through Oceanic Thinking." *Environment and Planning D: Society and Space* 33, no. 2 (2015): 247–64.

Stengers, Isabelle. "The Challenge of Ontological Politics." In de la Cadena and Blaser, *A World of Many Worlds,* 83–112.

Stengers, Isabelle. *In Catastrophic Times: Resisting the Coming Barbarism.* Translated by Andrew Goffey. London: Open Humanities Press, 2015.

Stengers, Isabelle. "Reclaiming Animism." *e-flux Journal,* no. 36 (July 2012). https://www.e-flux.com.

Stépanoff, Charles. "Human–Animal 'Joint Commitment' in a Reindeer

Herding System." *Journal of Ethnographic Theory* 2, no. 2 (2012): 287–312.

Stoler, Ann Laura, ed. *Imperial Debris: On Ruins and Ruination*. Durham, N.C.: Duke University Press, 2013.

Stoler, Ann Laura. " 'The Rot Remains': From Ruins to Ruination." In Stoler, *Imperial Debris*, 1–38.

Sukhova, C. A., T. S. Mostahova, A. C. Barashkova, D. V. Tumanova, and I. A. Elshina. "Demograficheskiye protsessy Respublike Sakha (Yakutiya)" [Demographic processes of the Sakha Republic (Yakutia)]. In *Territorial Aspect*. Yakutsk: Sakhaada, 2017.

Sumgin, Mikhail I. Vechnaya merzlota pochvy v predelakh SSSR [Permafrost soils in the USSR], 2nd ed. Moscow: Publishing House of the USSR Academy of Sciences, 1937.

Sussberg, Jason, and David Alvarado, dirs. *We Are as Gods*. New York: Structure Films, 2020.

Svenning, Jens-Christian, Pil B. M. Pedersen, C. Josh Donlan, Rasmus Ejrnæs, Søren Faurby, Mauro Galetti, Dennis M. Hansen, et al. "Science for a Wilder Anthropocene: Synthesis and Future Directions for Trophic Rewilding Research." *PNAS* 113, no. 4 (2016): 898–906.

Swyngedouw, Erik. "Apocalypse Forever? Post-political Populism and the Spectre of Climate Change." *Theory, Culture & Society* 27, nos. 2–3 (2010): 213–32.

Swyngedouw, Erik. "Apocalypse Now! Fear and Doomsday Pleasures." *Capitalism Nature Socialism* 24, no. 1 (2013): 9–18.

Szerszynski, Bronislaw. "The End of the End of Nature: The Anthropocene and the Fate of the Human." *Oxford Literary Review* 34, no. 2 (2012): 165–84.

Tarnocai, Charles, Josep G. Canadell, Edward A. G. Schuur, Peter Kuhry, Galina Mazhitova, and Sergey Zimov. "Soil Organic Carbon Pools in the Northern Circumpolar Permafrost Region." *Global Biogeochemical Cycles* 23, no. 2 (2009): 1–11.

Tatarchenko, Ksenia. "The Lena Is Worthy of Baikal: Defining Remoteness across the North and the East." *Technosphere Magazine*, November 6, 2018. https://technosphere-magazine.hkw.de.

TED. "Playlist: Is It Time for De-extinction?" Accessed July 7, 2020. https://www.ted.com.

Telesca, Jennifer E. *Red Gold: The Managed Extinction of the Giant Bluefin Tuna*. Minneapolis: University of Minnesota Press, 2020.

Thacker, Eugene. *The Global Genome: Biotechnology, Politics, and Culture*. Cambridge: MIT Press, 2005.

Thompson, Charis. *Making Parents: The Ontological Choreography of Reproductive Technologies*. Cambridge: MIT Press, 2007.

Tsing, Anna. "Unruly Edges: Mushrooms as Companion Species." *Environmental Humanities* 1, no. 1 (2012): 141–54.

Tsing, Anna Lowenhaupt. *The Mushroom at the End of the World: On the Possibility of Life in Capitalist Ruins.* Princeton, N.J.: Princeton University Press, 2015.

Turetsky, Merritt R., Benjamin W. Abbott, Miriam C. Jones, Katey Walter Anthony, David Olefeldt, Edward A. G. Schuur, Charles Koven, et al. "Permafrost Collapse Is Accelerating Carbon Release." *Nature* 569 (2019): 32–34.

Turner, Stephanie S. "Open-Ended Stories: Extinction Narratives in Genome Time." *Literature and Medicine* 26, no. 1 (Spring 2007): 55–82.

van Dooren, Thom. "Banking Seed: Use and Value in the Conservation of Agricultural Diversity." *Science as Culture* 18, no. 4 (2009): 373–95.

van Dooren, Thom. *Flight Ways: Life and Loss at the Edge of Extinction.* New York: Columbia University Press, 2014.

van Dooren, Thom, and Deborah Bird Rose. "Keeping Faith with the Dead: Mourning and De-extinction." *Australian Zoologist* 38, no. 3 (2017): 375–78.

Vasil'chuk, Yurij K., Johannes van der Plicht, Hogne Jungner, Eloni Sonninen, and Alla C. Vasil'chuk. "First Direct Dating of Late Pleistocene Ice-Wedges by AMS." *Earth and Planetary Science Letters* 179 (2000): 232–47.

Vera, Frans. "The Shifting Baseline Syndrome in Restoration Ecology." In *Restoration and History: The Search for a Usable Environmental Past,* edited by Marcus Hall, 98–110. New York: Routledge, 2010.

Verdery, Katherine. *The Political Lives of Dead Bodies: Reburial and Postsocialist Change.* New York: Columbia University Press, 1999.

Vernadsky, Vladimir I. "The Biosphere and the Noösphere." *American Scientist* 33, no. 1 (1945): 1–12.

Vitebsky, Piers. *Reindeer People: Living with Animals and Spirits in Siberia.* London: Harper Perennial, 2005.

Viveiros de Castro, Eduardo. "Cannibal Metaphysics: Amerindian Perspectivism." *Radical Philosophy* 182 (November/December 2013): 17–28.

Waldby, Catherine. *The Visible Human Project: Informatic Bodies and Posthuman Medicine.* London: Routledge, 2000.

Warde, Paul, Libby Robin, and Sverker Sörlin. "Stratigraphy for the Renaissance: Questions of Expertise for 'the Environment' and 'the Anthropocene.'" *Anthropocene Review* 4, no. 3 (2017): 246–58.

Waters, Colin N., Jan A. Zalasiewicz, Mark Williams, Michael A. Ellis, and Andrea M. Snelling. "A Stratigraphical Basis for the Anthropocene?" In *A Stratigraphical Basis for the Anthropocene,* edited by Colin N. Waters, Jan A. Zalasiewicz, Mark Williams, Michael A. Ellis, and Andrea M. Snelling, 1–21. London: Geological Society of London, 2014.

Watt-Cloutier, Siila. *The Right to Be Cold: One Woman's Story of Protecting Her Culture, the Arctic, and the Whole Planet.* Minneapolis: University of Minnesota Press, 2015.

Weisberger, Mindy. "Worms Frozen for 42,000 Years in Siberian Permafrost Wriggle to Life." Live Science, July 27, 2018. https://www.livescience.com.

Weiss, Sabrina. "The Climate Crisis Has Sparked a Siberian Mammoth Tusk Gold Rush." *Wired*, November 18, 2019. https://www.wired.co.uk.

Welch, Craig. "Some Arctic Ground No Longer Freezing—Even in Winter." *National Geographic*, August 20, 2018. https://www.nationalgeographic.com.

Wenzel, Jennifer. *Bulletproof: Afterlives of Anti-colonial Prophecy in South Africa and Beyond*. Chicago: University of Chicago Press, 2009.

Whatmore, Sarah. "Materialist Returns: Practising Cultural Geography in and for a More-Than-Human World." *Cultural Geographies* 13, no. 4 (2006): 600–609.

Whyte, Kyle Powys. "Indigenous Climate Change Studies: Indigenizing Futures, Decolonizing the Anthropocene." *English Language Notes* 55, nos. 1–2 (2017): 153–62.

Whyte, Kyle Powys. "White Allies, Let's Be Honest about Decolonization." *Yes!*, April 3, 2018. https://www.yesmagazine.org.

Winterton, Neil. "In Praise of (Some) Mavericks." *Clean Technologies and Environmental Policy* 7 (2007): 153–55.

Wolfe, Cary. "Foreword." In *Extinction Studies: Stories of Time, Death, and Generations*, edited by Deborah Bird Rose, Thom van Dooren, and Matthew Chrulew, vii–xvi. New York: Columbia University Press, 2017.

Worrall, Simon. "We Could Resurrect the Woolly Mammoth. Here's How." *National Geographic*, July 9, 2017. https://www.nationalgeographic.com.

Wray, Britt. *Rise of the Necrofauna: The Science, Ethics, and Risks of De-extinction*. Vancouver: Greystone Books, 2017.

Wrigley, Charlotte. "Ice and Ivory: The Cryopolitics of Mammoth De-extinction." *Journal of Political Ecology* 28, no. 1 (2021): 782–803.

Wrigley, Charlotte. "Nine Lives Down: Love, Loss, and Longing in Scottish Wildcat Conservation." *Environmental Humanities* 12, no. 1 (2020): 346–69.

WWF. *Living Planet Report—2018: Aiming Higher*. Edited by Monique Grooten and Rosamunde E. A. Almond. Gland, Switzerland: WWF, 2018.

Wynne-Jones, Sophie, Clara Clancy, George Holmes, Kieran O'Mahony, and Kim J. Ward. "Feral Political Ecologies? The Biopolitics, Temporalities and Spatialities of Rewilding." *Conservation and Society* 18, no. 2 (2020): 71–76.

Yamagata, Kazuo, Kouhei Nagai, Hiroshi Miyamoto, Masayuki Anzai, Hiromi Kato, Kei Miyamoto, Satoshi Kurosaka, et al. "Signs of Biological Activities of 28,000-Year-Old Mammoth Nuclei in Mouse Oocytes Visualized by Live-Cell Imaging." *Scientific Reports* 9 (March 2019). https://doi.org/10.1038/s41598-019-40546-1.

Yashina, Svetlana, Stanislav Gubin, Stanislav Maksimovich, Alexandra Yashina, Edith Gakhova, and David Gilichinsky. "Regeneration of Whole Fertile Plants from 30,000-y-Old Fruit Tissue Buried in Siberian Permafrost." *PNAS* 109, no. 10 (2012): 4008–13.

Yusoff, Kathryn. "Anthropogenesis: Origins and Endings in the Anthropocene." *Theory, Culture & Society* 33, no. 2 (2016): 3–28.

Yusoff, Kathryn. *A Billion Black Anthropocenes or None.* Minneapolis: University of Minnesota Press, 2018.

Yusoff, Kathryn. "Geologic Life: Prehistory, Climate, Futures in the Anthropocene." *Environment and Planning D: Society and Space* 31, no. 5 (2013): 779–95.

Yusoff, Kathryn. "Geologic Subjects: Nonhuman Origins, Geomorphic Aesthetics and the Art of Becoming Inhuman." *Cultural Geographies* 22, no. 3 (2015): 383–407.

Yusoff, Kathryn, and Jennifer Gabrys. "Climate Change and the Imagination." *WIREs Climate Change* 2, no. 4 (2011): 516–34.

Zalasiewicz, Jan, Colin N. Waters, and Mark Williams. "Human Bioturbation, and the Subterranean Landscape of the Anthropocene." *Anthropocene* 6 (June 2014): 3–9.

Zalasiewicz, Jan, Mark Williams, Richard Fortey, Alan Smith, Tiffany L. Barry, Angela L. Coe, Paul R. Bown, et al. "Stratigraphy of the Anthropocene." *Philosophical Transactions of the Royal Society A* 369, no. 1938 (2011): 1036–55.

Zalasiewicz, Jan, Mark Williams, Alan Smith, Tiffany L. Barry, Angela L. Coe, Paul R. Bown, Patrick Brenchley, et al. "Are We Now Living in the Anthropocene?" *GSA Today* 18, no. 2 (2008): 4–8.

Zimov, Sergey A. "Pleistocene Park: Return of the Mammoth's Ecosystem." *Science* 308, no. 5723 (2005): 796–98.

Zimov, Sergey A. "Wild Field Manifesto." Accessed May 22, 2019. https://justmeat.co.

Zimov, Sergey A., V. I. Chuprynin, A. P. Oreshko, F. Stuart Chapin III, J. F. Reynolds, and M. C. Chapin. "Steppe–Tundra Transition: A Herbivore-Driven Biome Shift at the End of the Pleistocene." *American Naturalist* 146, no. 5 (1995): 765–94.

Zimov, Sergey A., Edward A. G. Schuur, and F. Stuart Chapin III. "Permafrost and the Global Carbon Budget." *Science* 312, no. 5780 (2006): 1612–13.

Zimov, Sergey A., Nikita S. Zimov, and F. Stuart Chapin III. "The Past and Future of the Mammoth Steppe Ecosystem." In *Paleontology in Ecology and Conservation,* edited by Julien Louys, 193–225. Berlin: Springer, 2012.

Zimov, Sergey A., Nikita S. Zimov, Alexei N. Tikhonov, and F. Stuart Chapin III. "Mammoth Steppe: A High Productivity Phenomenon." *Quaternary Science Reviews* 57, no. 4 (2012): 26–45.

Index

Charlotte Wrigley is a postdoctoral research fellow in environmental history at the University of Stavanger, Norway. She was previously appointed as a postdoctoral fellow at the Laboratory for Environmental and Technological History in St. Petersburg, Russia, and conducted research on the Kola Peninsula and in the Sakha Republic.